BrightRED Study Guide

Curriculum for Excellence

N4

PHYSICS

Paul Van der Boon

First published in 2015 by:
Bright Red Publishing Ltd
1 Torphichen Street
Edinburgh
EH3 8HX

A CIP record for this book is available from the British Library

ISBN 978-1-906736-51-4

With thanks to:
PDQ Digital Media Solutions Ltd (layout), Clodagh Burke (edit)

Cover design and series book design by Caleb Rutherford – e i d e t i c

Acknowledgements
Every effort has been made to seek all copyright holders. If any have been overlooked, then Bright Red Publishing will be delighted to make the necessary arrangements.

Permission has been sought from all relevant copyright holders and Bright Red Publishing are grateful for the use of the following:

Charlie Nettle (CC BY 2.0)[1] (p 6); American Center Mumbai (CC BY-ND 2.0)[2] (p 6); Thomas Anderson (CC BY 2.0)[1] (p 6); Graeme Maclean (CC BY 2.0)[1] (p 6); xlibber (CC BY 2.0)[1] (p 7); Raining girl (CC BY 3.0)[3] (p 7); PSNH (CC BY-ND 2.0)[2] (p 7); Image licensed by Ingram Image (p 8); huubvanhughten (CC BY 2.0)[1] (p 10); Carlo Toffolo/Shutterstock.com (p 13); Capgros/freeimages.com (p 13); epSos.de (CC BY 2.0)[1] (p 13); Andrew Imanaka (CC BY 2.0)[1] (p 14); pelucco/iStock.com (p 14); Reuben Flounders (CC BY-ND 2.0)[2] (p 16); Lightyears.dk (CC BY-ND 2.0)[2] (p 16); serezniy/iStock.com (p 16); Tusumaru/Shutterstock.com (p 16); George Alexander Ishida Newman (CC BY 2.0)[1] (p 17); imagedb.com/Shutterstock.com (p 20); Geek3 (CC BY-SA 3.0)[4] (p 20); Palmer, Alfred T./Farm Security Administration (public domain) (p 20); Les Chatfield (CC BY 2.0)[1] (p 20); Image licensed by Ingram Image (p 21); Ambrozjo/freeimages.com (p 21); Fmarkos (public domain) (p 21); BrokenSphere (CC BY-SA 3.0)[4] (p 21); lolostock/iStock.com (p 21); Bork/Shutterstock.com (p 21); Timawe/iStock.com (p 25); Stu49/Shutterstock.com (p 25); Stu49/Shutterstock.com (p 25); herjua/iStock.com (p 24); eans/iStock.com (p 27); hisks/freeimages.com (p 27); Winai_Tepsuttinun/iStock.com (p 29); iphotographer/iStock.com (p 32); nilsz/iStock.com (p 32); aperturesound/Shutterstock.com (p 32); 4x6/iStock.com (p 32); AnonMoos based on image by User:Silje (CC BY-SA 3.0)[4] (p 32); Harvey Barrison (CC BY-SA 2.0)[5] (p 32); Mobilus In Mobili (CC BY-SA 2.0)[5] (p 32); Visivasnc/iStock.com (p 32); bhinddalenes (CC BY 2.0)[1] (p 33); Yuichiro Haga (CC BY 2.0)[1] (p 33); Dio5050/iStock.com (p 35); Gwenvidig/iStock.com (p 35); blueringmedia/iStock.com (p 38); photo5963/iStock.com (p 47); bogdanhoda/Shutterstock.com (p 47); amoceptum/iStock.com (p 50); Artur Synenko/Shutterstock.com (p 50); Christian Delbert/Shutterstock.com (p 50); Ilya Rabkin/Shutterstock.com (p 50); Stillwaterising (public domain) (p 50); kazina/iStock.com (p 50); Dan Foy (CC BY 2.0)[1] (p 50); Steven Lilley (CC BY-SA 2.0)[5] (p 50); Rashevskyi Viacheslav/Shutterstock.com (p 50); Ambrozjo/freeimages.com (p 50); cinezi/freeimages.com (p 50); threeart/iStock.com (p 53); danr13/iStock.com (p 52); Maximkostenko/iStock.com (p 52); Cary Bass (public domain) (p 58); Zdenko Zivkovic (CC BY 2.0)[1] (p 58); Denise Lett/Shutterstock.com (p 58); JohnCarnemolla/iStock.com (p 60); NH2501 (CC BY-SA 4.0)[6] (p 60); IAEA Imagebank (CC BY-SA 2.0)[5] (p 60); RIA Novosti archive, image #305008/Alexey Danichev (CC BY-SA 3.0)[4] (p 61); Ranglen/Shutterstock.com (p 61); William Warby (CC BY 2.0)[1] (p 63); Christopher Lofthouse (p 63); Jari Hindström/iStock.com (p 64); CityofStPete (CC BY-ND 2.0)[2] (p 66); Nic Redhead (CC BY-SA 2.0)[5] (p 66); Glen Wallace (CC BY-SA 2.0)[5] (p 66); Elliott Brown (CC BY 2.0)[1] (p 67); ssguy/Shutterstock.com (p 68); sound35/iStock.com (p 69); cheri131/iStock.com (p 69); Christopher Lofthouse (p 70); James Steidl/Shutterstock.com (p 71); ollyy/Shutterstock.com (p 71); mipan/iStock.com (p 72); Steve Jurvetson (CC BY 2.0)[1] (p 72); Republic of Korea (CC BY-SA 2.0)[5] (p 72); Robert Clemens (CC BY-ND 2.0)[2] (p 72); vichie81/Shutterstock.com (p 73); Maksim Toome/Shutterstock.com (p 73); FredFroese/iStock.com (p 74); katkov/iStock.com (p 74); Greg Goebel (CC BY-SA 2.0)[5] (p 76); NASA/JPL-Caltech/MSSS (p 76); NASA (p 77); NASA/JPL-Caltech-ESA/Hubble and Digitized Sky Survey 2 (p 77); NASA (p 77); 3DSculptor/iStock.com (p 77); Ken and Nyetta (CC BY 2.0)[1] (p 77); Ken and Nyetta (CC BY 2.0)[1] (p 78); isskh/Shutterstock.com (p 83); Galushko Sergey/Shutterstock.com (p 83); Lucky Business/Shutterstock.com (p 83); Serg64/Shutterstock.com (p 83); nikkytok/Shutterstock.com (p 83); Levent Konuk/Shutterstock.com (p 83); NASA and The Hubble Heritage Team (STScI/AURA) (p 84); NASA/Swift/Stefan Immler (GSFC) and Erin Grand (UMCP) (p 85); Harman Smith and Laura Generosa/NASA (p 85); ESO/Mario Nonino, Piero Rosati and the ESO GOODS Team (CC BY 4.0)[7] (p 85); fotokostic/iStock.com (p 86); Tony Hisgett (CC BY 2.0)[1] (p 86); IPGGutenbergUKLtd/iStock.com (p 86); redjar (CC BY-SA 2.0)[5] (p 88); Port of San Diego (CC BY 2.0)[1] (p 88); Christopher Lofthouse (p 88).

(CC BY 2.0)[1]	http://creativecommons.org/licenses/by-sa/2.0/
(CC BY-ND 2.0)[2]	http://creativecommons.org/licenses/by-nd/2.0/
(CC BY 3.0)[3]	http://creativecommons.org/licenses/by/3.0/
(CC BY-SA 3.0)[4]	http://creativecommons.org/licenses/by-sa/3.0/
(CC BY-SA 2.0)[5]	http://creativecommons.org/licenses/by-sa/2.0/
(CC BY-SA 4.0)[6]	https://creativecommons.org/licenses/by-sa/4.0/
(CC BY 4.0)[7]	http://creativecommons.org/licenses/by/4.0/

Printed and bound in the UK by Ashford Colour Ltd

CONTENTS

BRIGHT RED STUDY GUIDE: NATIONAL 4 PHYSICS

1 ELECTRICITY AND ENERGY

2 WAVES AND RADIATION

3 DYNAMICS AND SPACE

4 ADDED VALUE UNIT

5 ANSWERS

INTRODUCING NATIONAL 4 PHYSICS

Studying the N4 Physics course will provide you with the opportunity to develop an interest in, and an understanding of, the world we live in. At the same time your skills of communication and critical thinking will improve. You will also develop an understanding of the role of physics in the scientific issues that affect society.

Success in the National 4 Physics course provides the skills, knowledge and understanding required to allow progression to the next level: the National 5 Physics course.

ABOUT THE NATIONAL 4 PHYSICS COURSE

The National 4 Physics course consists of four units. Three of these units specify the suggested coursework; however, there is considerable flexibility in the National 4 Physics course in that there is no specified mandatory content. Knowing your abilities first hand, your teacher or lecturer will decide the most appropriate ways to cover the Key Areas of the course. The Key Areas covered during the course are assessed by your teachers or lecturers, who will devise their own assessment programmes. They enjoy some flexibility in how they conduct these assessments to give maximum variety and interest to the assessment process. Sometimes they will be able to assess several outcomes at once; at other times, you will be given discrete tasks to perform, each assessing one skill.

The Key Areas are:

Unit 1: Electricity and energy

- generation of electricity
- electrical power
- electromagnetism
- practical electrical and electronic circuits
- gas laws and the kinetic model

Unit 2: Waves and radiation

- wave characteristics
- sound
- the electromagnetic spectrum
- nuclear radiation

Unit 3: Dynamics and space

- speed and acceleration
- relationship between forces, motion and energy
- satellites
- cosmology

Added Value Unit

The fourth unit of the course is the Added Value Unit. In this unit, you will draw on and apply the skills and knowledge you have learned during the course. The Added Value Unit is assessed through an assignment.

The assignment will be an in-depth study of a topical issue from a key area of the course chosen by you in agreement with your teacher or lecturer. The assignment will be assessed by your teacher or lecturer. The assignment is carried out under controlled conditions.

To prepare for this assessment you will choose and research or investigate an appropriate topic, focusing particularly on its applications and impact on society or the environment. Following this research, you will then process the information.

During the assessment you will present evidence of your ability to:

- describe the issue being investigated and its relevance to the environment/society;
- select appropriate information from at least two relevant sources;
- present information appropriately;
- describe the physics of the issue and its impact on the environment/society.

ABOUT THIS STUDY GUIDE

The National 4 Physics course consists of three units of suggested content.

The **Key Areas** of each unit are discussed in detail in this Study Guide, using explanations, pictures, diagrams, drawings and examples. Helpful hints are provided throughout the book in the **Don't forget** features, while there are plenty of opportunities to practise applying your knowledge through **Things to do and think about**. Some of the skills you will be expected to demonstrate are also covered in the book. These may be in the relevant Key Areas or covered in separate questions.

There are many opportunities to test yourself to check whether you have understood the ideas and explanations.

Questions

Quick questions follow each Key Area to test your understanding and to let you practise some of the ideas and techniques.

Extended questions feature at the end of each unit. These questions are designed to help you master the problem-solving techniques and skills required to complete the assessments. Answers to both the quick questions and extended questions are provided towards the end of this Study Guide.

ELECTRICITY AND ENERGY

GENERATION OF ELECTRICITY: WHERE OUR ELECTRICITY COMES FROM 1

Different ways of generating electricity.

People need electricity every day. Countries require vast amounts of electrical energy for people to use – in homes, for transport and in industry.

Most countries use a variety of sources of energy for the generation of electricity. These different sources of energy are converted into electrical energy. This helps countries to avoid reliance on a single source of energy that may become scarce or cause pollution.

In Scotland, power stations use a variety of energy sources to generate electricity.

ONLINE

You can monitor the demand and supply for electricity in the UK on this website: http://www.gridwatch.templar.co.uk/

THERMAL POWER STATIONS

In a thermal power station, a fossil fuel (usually coal or gas) containing **chemical energy** is burned to produce **heat energy**. This heat energy is used to boil water and the steam produced turns the blades of a turbine. The energy change here is from **heat energy** to **kinetic energy**. The rotating turbine is connected to a generator that produces electricity. The energy change in the generator is from **kinetic energy** to **electrical energy**.

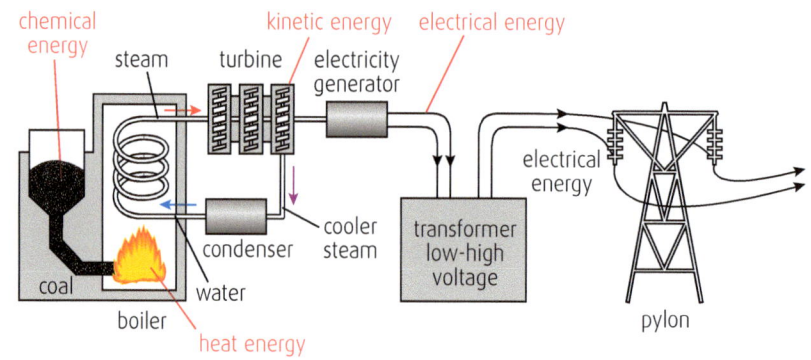

Transformations of energy in a thermal power station.

COAL-FIRED POWER STATIONS

The United Kingdom still relies on some coal-fired power stations to supply electrical energy. A disadvantage of coal-fired power stations is the pollution from gases released into the atmosphere. The main pollutant is carbon dioxide. Carbon dioxide gas is thought to be a major contributor to global warming.

The coal-fired power station at Longannet in Fife, one of the largest in Europe, will soon be closed.

TASK

Carry out research into the different types of pollutants produced by coal-fired power stations. Write down a list of what they are and what effect they have on people and our planet. Find out whether they are being replaced by other less polluting power stations.

DON'T FORGET

Biomass power stations use wood as an alternative to coal.

HYDROELECTRIC POWER STATIONS

Hydroelectric power stations convert the gravitational potential energy of water stored behind a dam into kinetic energy as it runs down pipes. This energy, in turn, is then converted into electrical energy by generators when the water rotates turbines connected to the generator.

Hydroelectric power stations can only be constructed in suitable areas, such as mountainous regions or on land that is at a higher level than the surrounding countryside. There is little pollution from these power stations.

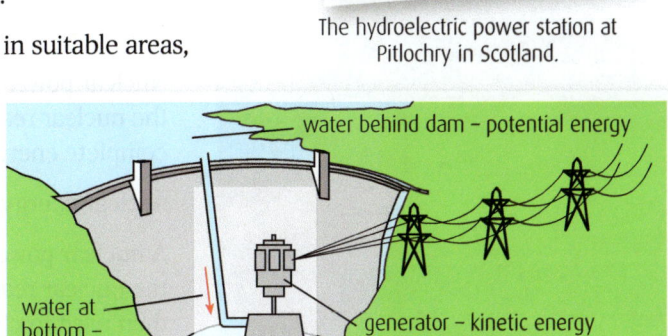

The hydroelectric power station at Pitlochry in Scotland.

PUMPED HYDROELECTRIC SCHEMES

A pumped hydroelectric power station pumps water back up to the reservoir, using cheaper off-peak electricity at night when there is low demand. This water flows back down to the generator during the daytime, generating electricity that is sold to customers at the peak (higher) rate.

water behind dam – potential energy

water at bottom – kinetic energy

generator – kinetic energy to electrical energy

Hydroelectric power stations use fast moving water to turn the generator.

BIOMASS POWER STATIONS

Biomass power stations burn renewable fuel sources, such as wood, straw or poultry litter, to produce heat energy. The fuel sources are managed to produce a sustainable supply. However, replanted forests take time to grow and require very large areas of land to make the process sustainable. The power output from biomass power stations in Scotland is currently fairly low compared with that from power stations using non-renewable sources.

top reservoir

pumps/ turbines

lower reservoir

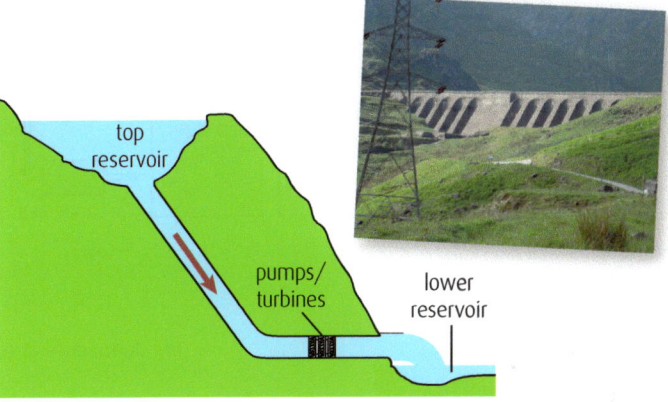

The reservoir supplying Ben Cruachan, a hydroelectric power station in the West Highlands of Scotland.

Biomass power stations burn wood from sustainable sources to produce electricity.

THINGS TO DO AND THINK ABOUT

1. For coal-fired power stations:
   (a) What energy change takes place: (i) in the boiler; and (ii) in the electricity generator?
   (b) What is the main polluting gas produced by coal-fired power stations?
   (c) What environmental problem do many scientists think is caused by this gas?

2. For hydroelectric power stations:
   (a) What energy change takes place: (i) when water flows from the dam to the electricity generator below; and (ii) in the electricity generator?
   (b) State an advantage of hydroelectric power stations compared with coal-fired power stations.
   (c) Where are hydroelectric power stations most commonly found?
   (d) Give one advantage of a pumped hydroelectric station over a normal hydroelectric power station.

GENERATION OF ELECTRICITY: WHERE OUR ELECTRICITY COMES FROM 2

A nuclear power station with two reactors.

NUCLEAR POWER STATIONS

Nuclear power stations use nuclear fuel to produce **heat energy** in the nuclear reactor. Steam is then used to turn the generator. The complete energy changes are:

nuclear energy → heat energy → kinetic energy → electrical energy

A nuclear power station produces heat energy inside the reactor via the nuclear reactions that take place in the fuel. This heat energy is then used to change water into steam and to drive turbines attached to a generator, as in thermal power stations.

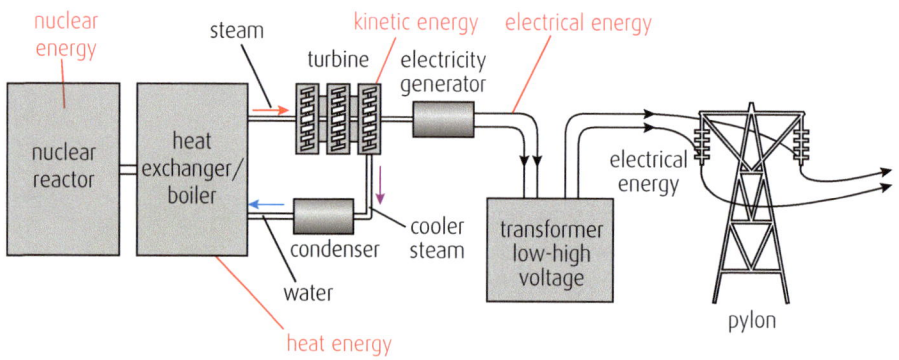

Transformations of energy in a nuclear power station.

Nuclear fission

In nuclear reactors, a uranium nucleus is bombarded by a neutron and splits into two smaller nuclei and more neutrons. This splitting of the nucleus is called **nuclear fission**. There is more kinetic energy after the fission process than before and this is how heat is produced.

The extra neutrons released go on to cause more fission and more heat. This is called a **chain reaction**.

Nuclear fission is a different process to the burning of fossil fuels and equal masses of coal and nuclear fuel release very different amounts of energy.

The amount of electrical energy produced in a power station compared with the energy released by the fuel is a measure of the **efficiency** of the generation process.

Nuclear power stations produce very little carbon dioxide gas and are not thought to contribute to global

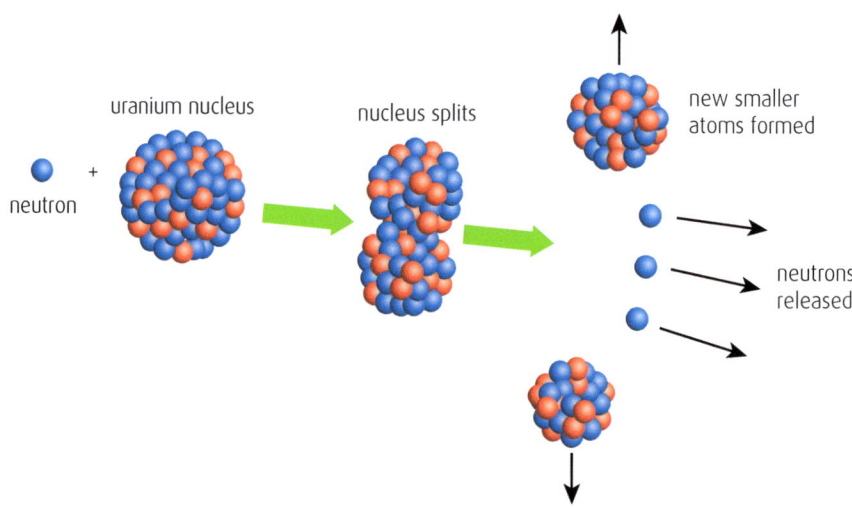

Fission process for a uranium nucleus used as a nuclear fuel.

warming. Also, compared with coal-fired power stations, the mass of nuclear fuel required to produce the same amount of electrical energy is much smaller than that of coal. However, the used nuclear fuel (known as nuclear waste) is very dangerous and many precautions must be taken to protect both workers and the public.

When the nuclear fuel has reached the end of its useful life, it must be stored carefully and securely for many years until the radiation levels are safe. This storage is very expensive. There have been accidents at some nuclear power stations that have caused long-lasting and damaging effects.

fusion process

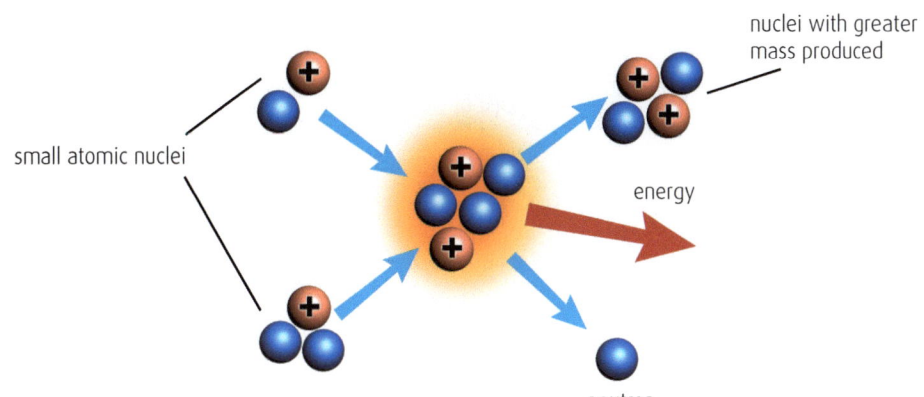

Nuclear fusion: smaller atomic nuclei collide to produce a nucleus with a greater mass.

Nuclear fusion

Nuclear fusion occurs when the nuclei of two atoms combine to form a nucleus with a larger mass. When this happens, a large amount of energy is released. This process is the source of energy in stars, including our Sun.

No nuclear waste is produced in nuclear fusion reactions. Scientists are working on the construction of experimental nuclear fusion reactors for use in nuclear power stations.

THINGS TO DO AND THINK ABOUT

1. Some of the stages in a nuclear power station are shown in the following diagram:

At which stage is the main energy transformation: (a) kinetic → electrical energy; and (b) nuclear → heat energy?

2. Explain how heat energy is produced inside a nuclear reactor.

3. What are the main advantages of nuclear power stations compared with coal-fired power stations?

4 What is the main disadvantage of nuclear power generation?

GENERATION OF ELECTRICITY: WHERE OUR ELECTRICITY COMES FROM 3

HOW ELECTRICITY IS GENERATED IN POWER STATIONS

It is possible to create a voltage and current in a coil of wire if a magnet is moved towards (or away from) the coil.

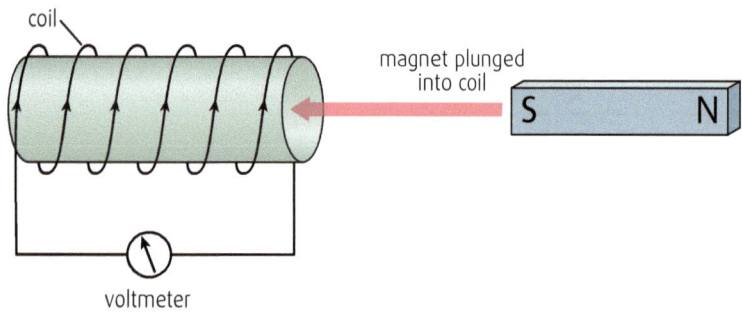

Generation of an alternating (a.c.) voltage.

A current can also be generated if the coil is moving and the magnet is stationary. The main parts of an a.c. generator are:

- a coil
- an iron core
- a magnet

When a magnet is moved inside a coil of wire (or a coil of wire is moved around the magnet), a voltage is produced.

The **size** of the a.c. **voltage** produced depends on the **strength** of the magnetic field, the **speed** of movement of the magnet or coil and the number of **turns** of wire on the coil.

A bicycle dynamo is an a.c. generator that converts kinetic energy into electrical energy to light the lamps. When the bicycle moves, the wheel turns the magnet inside the iron core and produces a voltage in the coil of wire.

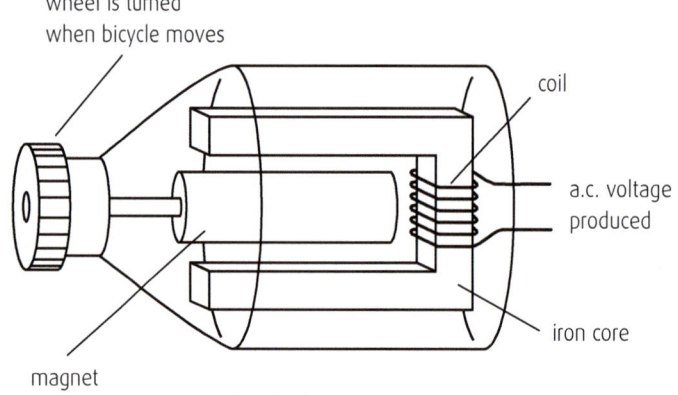

How a bicycle dynamo is constructed.

Using permanent magnets is impractical in power stations because they would have to be very large and powerful. Coils of wire are used instead of permanent magnets.

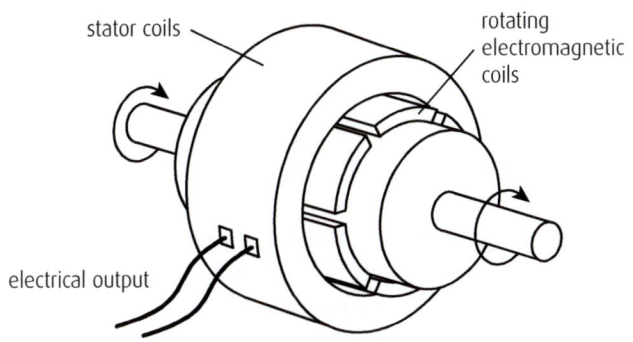

A power station uses an a.c. generator to convert kinetic energy into electrical energy.

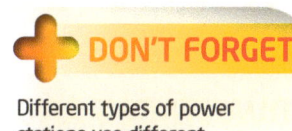

DON'T FORGET

Different types of power stations use different energy sources.

 TASK   Quick questions

1. Describe how a voltage can be produced using a coil of wire.
2. How can this voltage be increased?

THINGS TO DO AND THINK ABOUT

1. Some accidents have occurred that have damaged nuclear power stations. Carry out research into the two most recent major accidents at nuclear power stations in the world. Try to find out the answers to the following:
   - What was the cause of each accident?
   - What happened to people and property in the immediate area?
   - What happened to people and property in more distant places?
   - Were all the problems cleared up after the accident?
   - What do supporters of nuclear power production say to defend their point of view?
   - What do people who object to nuclear power production say to defend their point of view?
   - Are there any countries that have stopped using nuclear power or who have stopped building new nuclear power stations?

2. Carry out an investigation into the generation of electricity using a coil of wire and a magnet to show how the number of turns of wire in the coil affects the size of the voltage produced. Use apparatus similar to that shown in the diagram at the start of this section. Think about what apparatus you will need.

   In your notebook, write down how you will change the number of turns in the coil of wire and how you will keep the investigation fair (especially when moving the magnet).

   Collect your apparatus and construct a table in your notebook to record your results.

   Write down a statement about what you find.

GENERATION OF ELECTRICITY: WHERE OUR ELECTRICITY COMES FROM 4

ADVANTAGES AND DISADVANTAGES OF DIFFERENT ENERGY SOURCES

There is a huge and increasing demand for electrical energy in most countries of the world. The input energy sources used by many countries cause atmospheric pollution. Many scientists agree that this pollution is causing many problems for the planet. They believe that this pollution should be reduced, where possible, by using less polluting sources of input energy. Different energy sources have both advantages and disadvantages. Renewable energy sources are being developed and used to supply some of our energy needs. Many countries still rely on polluting energy sources such as coal or gas for use in power stations.

Renewable sources of energy	Advantages	Disadvantages
Solar	Will not run out, non-polluting, free at source	Poor output in cooler countries Requires large area of solar cells to produce significant output
Geothermal	Causes little pollution Once the water is used, it can be pumped back into the ground to be re-used	Limited number of sites available Difficult to convert energy produced
Water (hydro/waves/tides)	Causes little pollution Provides a 'constant' source of energy The water returns to its natural origin	Requires mountains (hydroelectric) or large areas of sea (for wave power)
Wind	Causes little pollution Can supply electrical energy in remote locations.	Wind supply is variable
Biomass	Requires little technology to set up (useful in poorer countries)	Slow to generate energy, limited output, requires large areas of land to grow energy source

Non-renewable sources of energy	Advantages	Disadvantages
Coal	Long-term supply still available in some countries	Produces carbon dioxide gas, causes pollution, dirty and can cause damage to surrounding areas
Oil	Used as main source of fuel and in manufactured products	Produces carbon dioxide gas, limited supplies available
Gas	Clean and easy to distribute	Produces carbon dioxide gas, limited supply available Controversial new methods of extraction (e.g. fracking)
Nuclear (fission, using uranium)	Only small amounts of fuel source are required, long lifetime of supply	Waste material remains hazardous for long periods of time Accidents can cause severe pollution
Nuclear (fusion)	Plentiful supply of input material, safe	Theoretical form of energy conversion still under development

FUTURE SUSTAINABLE ELECTRICITY SUPPLY: MICROGENERATION

The 'microgeneration' of electricity means the use of smaller, local sources of renewable energy to generate electricity. This type of electricity generation may help to supply some of our energy needs in the future. Any of the electricity generated that is not required by the householder is returned to the National Grid and used elsewhere. The householder is credited with the value of this electricity, thus reducing the household electricity bill.

Solar (photovoltaic) cells

Solar (or photovoltaic) cells can be linked together into larger arrays and used in domestic situations to provide household electricity. The electricity generated from solar energy is connected to the electricity supply for the house.

Solar cells on the roof of a house.

Water turbines

Water turbines attached to a generator can be installed in small rivers or streams and used to generate electricity.

Small wind turbines

Small wind turbines can be used in appropriate locations to generate electricity. Such turbines are common in rural areas or on islands where there are suitable wind conditions and enough space to erect the turbines.

A water turbine for use in small rivers or streams.

⚙ TASK Microgeneration

Carry out research into these different of methods of microgeneration. Search online for 'Small hydro microgeneration of electricity'.

Write a description of a microgenerating system that could use a source of renewable energy in your neighbourhood.

Small wind turbines can be put up near houses.

⚙ TASK Quick question

1. Solar cells are used on many houses to produce electrical energy.
   (a) State one **advantage** of using the Sun as a source of energy.
   (b) State one **disadvantage** of using the Sun as a source of energy.

DON'T FORGET

All energy sources have advantages and disadvantages.

THINGS TO DO AND THINK ABOUT

Research the ITER fusion project, where work is being carried out to construct a reactor for the generation of electrical energy from **nuclear fusion**. Find out the advantages and disadvantages of nuclear fusion.

Investigate the factors that affect the output voltage produced by a **solar cell**. Think about the intensity of the light or the size (area) of the solar cell, or the distance of the light source from the solar cell.

Investigate how **tides** and **wave energy** can be harnessed as the input energy to generate electricity.

Try to find out:

- For tidal energy, has this technology actually been installed anywhere to produce a permanent supply of electricity for the National Grid?
- For wave energy, has this technology actually been installed anywhere to produce a permanent supply of electricity for the National Grid?
- Where are the most suitable places around Scotland for power stations to be built for these types of electricity generation?

DISTRIBUTION OF ELECTRICITY

Electricity needs to be distributed from the power stations where it is generated.

THE NATIONAL GRID

Once electricity has been generated, it needs to be distributed to cities, towns and factories all over the country.

The distribution system in the UK is known as the **National Grid**. The National Grid is responsible for transferring the electrical energy generated in power stations to where it is needed across the country.

REDUCING WASTAGE WHEN TRANSFERRING ELECTRICAL ENERGY

Electrical energy is transferred from place to place in overhead cables or wires. The current in the wires causes heat energy to be produced. This heat energy reduces the amount of electrical energy reaching the destination. Too much energy lost as heat is a waste of useful energy. The operators of the National Grid try to minimise the amount of energy wasted in this way. This improves the **efficiency** of the National Grid.

TRANSFORMERS

One way of reducing this wasted energy is to use **transformers** to allow energy from the power station to be transmitted across the country at very high voltages, up to 440 000 volts (440 kV). The voltage is then reduced to safer levels when it reaches towns and cities.

Transformers use the effect of magnetic fields to increase (step-up) or decrease (step-down) the value of an a.c. input voltage.

Transformers have an input coil called a **primary coil** and an output coil called a **secondary coil**. An iron core links the two coils. The input and output voltages are related by the relationship:

$$\frac{V_s}{V_p} = \frac{n_s}{n_p}$$

where V_s = the secondary (output) voltage, V_p = the primary (input) voltage, n_s = the number of turns of wire on the secondary coil and n_p = the number of turns of wire on the primary coil.

Transformers are used whenever a high voltage needs to be stepped down or a low voltage needs to be stepped up.

Overhead transmission lines (electricity pylons) are widely used to transfer electrical energy because burying cables underground over long distances is far more expensive.

ONLINE

Watch the YouTube video about reducing energy loss from power stations at http://goo.gl/kvnwuv/

Transformers are used to change the size of an a.c. voltage.

The voltage from the power station generators is usually 25 000 or 33 000 V. Transformers are used to change this input voltage into output voltages as high as 400 000 V.

This high voltage is transmitted by overhead cables supported by pylons. On reaching a town or city, the voltage is reduced by another transformer.

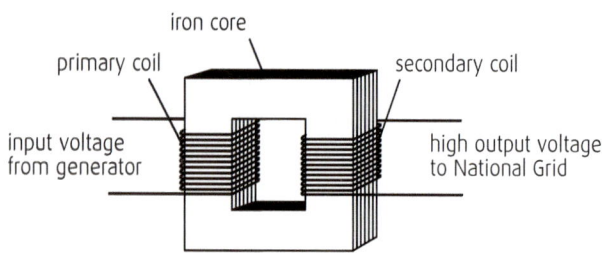

Diagram of a simple transformer.

iron core
primary coil
secondary coil
input voltage from generator
high output voltage to National Grid

EXAMPLE

A 5 V battery in a mobile phone is recharged from the mains using a charger containing a step-down transformer. The transformer consists of three parts: a **core**, a **primary coil** and a **secondary coil**.

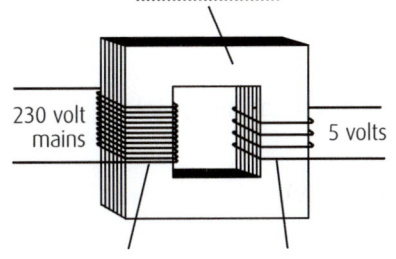

(a) Label these parts on the diagram.

(b) There are 11 500 turns on the primary coil of the transformer.

Calculate the number of turns on the secondary coil.

(c) Explain why a transformer cannot be used to step-down the voltage from a battery.

ANSWER

(a)

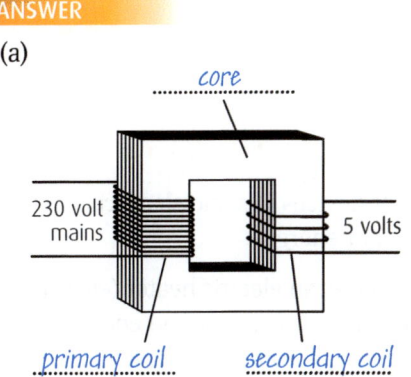

(b) $\dfrac{V_s}{V_p} = \dfrac{n_s}{n_p}$

$\dfrac{5}{230} = \dfrac{n_s}{n_p} = \dfrac{n_s}{11\,500}$

$n_s = \dfrac{5 \times 11\,500}{230} = 250$ turns

(c) Transformers only operate on an alternating current; a battery supplies a direct current.

 TASK   **Quick questions**

1. How is electrical energy transported across the country to the towns and cities where it is needed?

2. Why are transformers used in the process of transferring electrical energy across the country?

3. A laptop computer uses a charging unit that contains a transformer. The transformer reduces the mains voltage of 230 V to a safe voltage to recharge the laptop battery. The primary coil of the transformer has 16 560 turns of wire and the secondary coil has 2160 turns of wire.

   Calculate the output voltage of the charging unit that charges the laptop battery.

 **DON'T FORGET**

Transformers are used to step-up or step-down a.c. voltages.

THINGS TO DO AND THINK ABOUT

Overhead transmission lines carried by pylons are much cheaper than underground cables buried beneath roads. However, people sometimes ask for underground cables to be used in some areas.

- Find out about any recent campaigns to have electricity cables buried underground instead of suspended by pylons.
- Find out the reasons why people want cables to be buried.
- Find out what makes burying cables more expensive.
- Find out which are safest: buried or overhead cables.

ELECTRICAL POWER 1

POWER, ENERGY AND TIME

Appliances or devices in some electrical circuits convert electrical energy into another form of energy.

Examples include: an electric heater (electrical → heat energy), a loudspeaker (electrical → sound energy) or a lamp (electrical → light energy).

The rate at which energy is converted is known as the **power** of the component. Power is measured in **watts** (W). One watt is equal to one **joule per second**. If a 5 W LED lamp is switched on, it will convert (or dissipate) five joules of energy per second (5 Js⁻¹).

Different household appliances have different **power ratings**.

Appliances that involve the conversion of electrical energy into heat energy usually have higher power ratings than other appliances.

A 500 W bathroom towel radiator converts 500 J of electrical energy into heat each second.

Lamp

A 40 W LED television converts 40 J of energy each second.

⚙ TASK Quick questions

1. What is the unit of electrical power?

2. Which type of domestic electrical appliance usually has the largest power rating?

POWER CALCULATIONS

Power can be calculated using the relationship:

$P = \dfrac{E}{t}$ where P is the power in watts (W), E is the energy in joules (J) and t is the time in seconds (s).

Loudspeaker

EXAMPLE

A 480 W hairdryer is switched on for eight minutes.

Calculate the energy used.

ANSWER

$$P = \frac{E}{t}$$

First, we have to convert eight minutes into seconds:

8 minutes = (8 × 60) seconds

$$480 = \frac{E}{(8 \times 60)}$$

$$E = 480 \times 8 \times 60 = 230\ 400\ J$$

EXAMPLE

A 1200 W heater dissipated 3 240 000 J of heat energy.

How long was it switched on for?

ANSWER

$$P = \frac{E}{t}$$

$$1200 = \frac{3\ 240\ 000}{t}$$

$$t = \frac{3\ 240\ 000}{1200} = 2700\ s = 45\ minutes$$

SAVING ENERGY WHEN USING ELECTRICAL APPLIANCES

Appliances such as cookers, kettles, irons and toasters are deliberately designed to convert electrical energy into heat energy (to heat food, boil water, iron clothes and make toast).

Other appliances – for example, a washing machine or a dishwasher – produce heat energy as part of their normal operation (to heat water for cleaning).

When appliances are in use, some of the heat energy produced escapes to the environment (heating up the surrounding air or the actual machine) and is wasted.

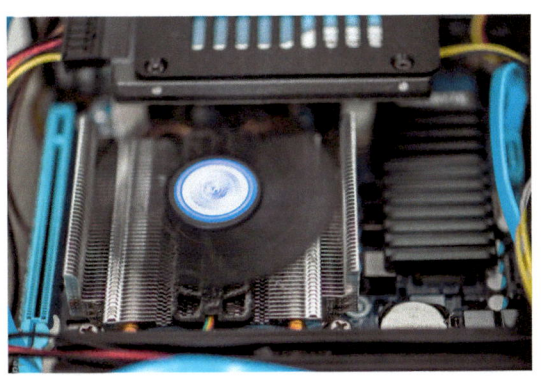

Computer microprocessor with cooling fan attached to prevent overheating.

This happens even when the appliance was not designed to produce heat energy. We say that this heat energy is wasted because the appliance was not designed to produce it.

For example, computers become warm when in use because the microprocessor inside the computer produces heat energy as it operates. This wasted heat energy adds to the overall energy required to operate the computer.

> **DON'T FORGET**
>
> Electrical appliances have a 'rating plate' that states the power of the appliance.

THINGS TO DO AND THINK ABOUT

1. A 3000 W microwave oven was switched on for two minutes.

   How much energy was used in the process?

2. A computer game station used 1 008 000 J of electrical energy during two hours of gaming.

   Calculate the power rating of the computer game station.

ELECTRICAL POWER 2

EFFICIENCY OF ELECTRICITY GENERATION

When an energy source is converted into electrical energy in a power station, there is always more energy required at the input stage than is finally produced as useful electrical energy at the output stage.

This is because energy is lost at different stages during the conversion process.

For example, in a coal-fired power station, the coal is burned to produce heat energy to change water into steam.

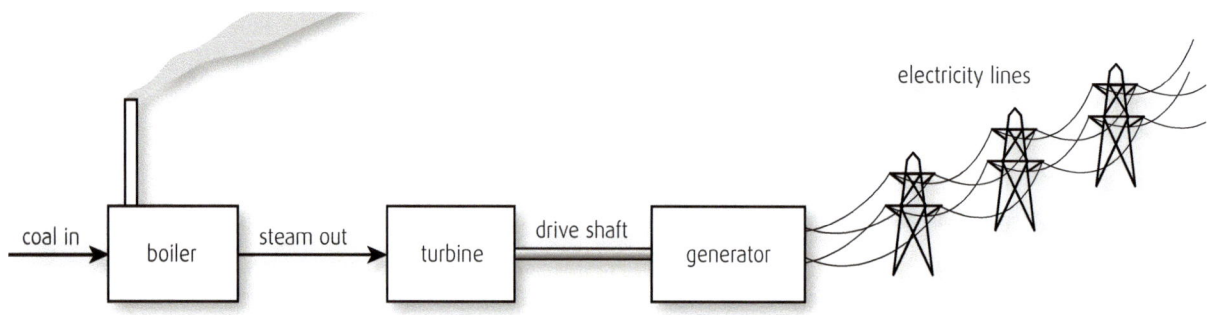

Energy conversion processes in a coal-fired power station.

During this process, some heat energy escapes to the atmosphere or is used to heat up the boiler containing the water.

As the steam travels to the turbines, it loses heat energy as the pipes heat up. There are also frictional forces inside the turbine and generator, meaning that less energy is available to generate electricity.

These energy 'losses' cause the electrical energy output to be less than the input energy.

useful energy out $<$ total input energy

(electrical energy)   less than

This means that power stations are not completely efficient – that is, they are less than 100% efficient.

Efficiency is calculated using the following relationships:

For **energy** calculations:

$$\% \text{ efficiency} = \frac{\text{useful } E_o}{E_i} \times 100$$

where E_o is the energy output and E_i is the energy input.

For **power** calculations:

$$\% \text{ efficiency} = \frac{\text{useful } P_o}{P_i} \times 100$$

where P_o = power output and P_i = power input.

EXAMPLE

A power station produces 500 MJ of heat energy.

The generator connected to the turbine produces 350 MJ of electrical energy.

Calculate the efficiency of this generator.

ANSWER

$$\text{Percentage efficiency} = \frac{\text{useful energy out}}{\text{total energy in}} \times 100$$

$$= \left(\frac{350 \times 10^6}{500 \times 10^6}\right) \times 100$$

$$= 70\%$$

Efficiency calculations can be applied to any process where there is an input of energy that is converted into a useful output of energy.

EXAMPLE

A lift transports five people from the ground floor of a building to an upper level.

The passengers gain a total of 69 600 J of potential energy.

The lift motor uses 120 000 J of electrical energy to raise both the lift and passengers.

Calculate the efficiency of the lift.

ANSWER

$$\text{Percentage efficiency} = \frac{\text{useful energy out}}{\text{total energy in}} \times 100$$

$$= \frac{69\ 600}{120\ 000} \times 100$$

$$= 58\%$$

TASK   Quick questions

1. Calculate the efficiency of an 18 W light bulb that produces light of power 7 W.

2. It has been calculated for some cars that, for every 1400 J of chemical energy of fuel used, only 378 J of useful kinetic energy is produced.

   Calculate the efficiency of this process.

DON'T FORGET

Electrical power is a measure of the energy used each second.

THINGS TO DO AND THINK ABOUT

1. Find the power ratings of some common appliances in your home.
   (a) Construct a table of your results.
   (b) Which of the appliances in your table converts the most joules of energy each second?

2. Research the government website: https://goo.gl/89HgMM to find out how appliances are labelled with a colour code to identify how much energy they waste when the appliance is in use.

   Find this rating safely on a domestic appliance in your home to determine its energy efficiency.

ELECTROMAGNETISM

MAGNETIC FIELDS

Permanent magnets

All magnets have a **north** and a **south pole**.

There is an **attracting** force between **different** poles.

There is a **repelling** force between the **same** poles.

A **magnetic field** exists around the magnet.

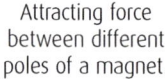

Attracting force between different poles of a magnet.

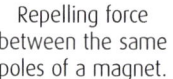

Repelling force between the same poles of a magnet.

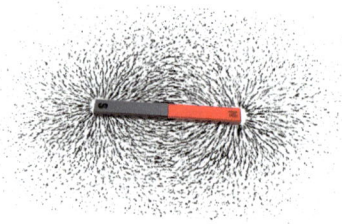

Iron particles showing the magnetic field pattern around a magnet.

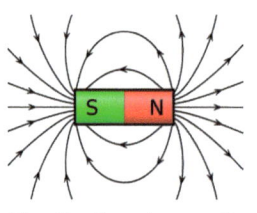

The direction of magnetic field lines is north to south (indicated by the direction of the arrows).

Some metals, when placed within this magnetic field, experience a force.

The magnetic field is invisible, but it can be visualised by sprinkling fine particles of iron on a clear plastic sheet covering a magnet.

The magnetic field pattern is represented by drawing **magnetic field lines**.

TASK   Quick questions 1

1. How can a magnetic field be represented?
2. In which direction are the arrows on magnetic field lines drawn?
3. Where is the magnetic field of a permanent bar magnet the strongest?

The magnetic field is stronger where the magnetic field lines are closer together (this happens at the poles of the magnet). This explains why magnetic materials are attracted to the poles of the magnet where the attracting force is greater.

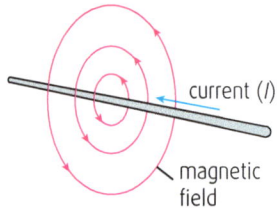

current (*I*)

magnetic field

Magnetic field around a wire carrying an electrical current.

Electromagnets

When there is a current in a wire, a magnetic field is produced around the wire.

The greater the current, then the stronger the magnetic field.

If the wire is looped into a coil (sometimes called a **solenoid**), then the magnetic field is stronger.

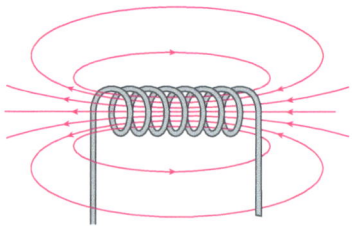

A wire can be looped into a coil or solenoid.

The magnetic field pattern is now the same as that of a permanent magnet. This means that the solenoid acts in the same way as an ordinary magnet.

When the solenoid is wound around an **iron core**, the magnetic field strength increases. This is known as an **electromagnet**.

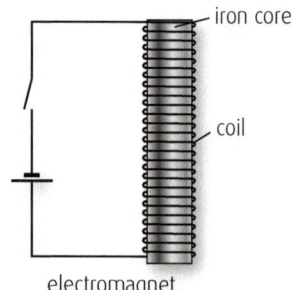

iron core

coil

electromagnet

A solenoid wrapped around an iron core is an electromagnet.

This electric bell consists of an electromagnet that is repeatedly energised to attract a metal hammer, which strikes the bell.

When the power is switched on, there is a current in the coil and a strong magnetic field is produced. The electromagnet will now attract magnetic materials. When the power is switched off, the magnetic field disappears. Electromagnets have many useful applications because the magnetic field can be switched on and off when required.

A crane with an electromagnet to lift and move magnetic materials.

MAGNETS OR ELECTROMAGNETS?

Some useful applications of magnetism require either permanent magnets or electromagnets.

Application	Picture	Circuit diagram or symbol	Type of magnet	Reason
Electric motor (d.c.)		—(M)—	Permanent magnet	Development of new stronger permanent magnets allows d.c. motors to be lighter and more controllable
Loudspeaker			Permanent magnet	Permanent magnets produce the constant magnetic field required for operation of the loudspeaker
Relay		coil	Electromagnet	The relay can be switched on and off when required, allowing external circuits to be operated
Magnetic door catch for fire doors			Electromagnet	Allows door to be held open during daytime and closed automatically when fire alarm is activated
Fridge door seal			Permanent magnet	Permanent magnetic strip allows door to be sealed when shut; does not require a power supply
MRI scanner			Electromagnet	Only electromagnets can produce the extremely strong magnetic fields required

TASK Quick questions 2

1. What is produced around a wire when an electric current flows through the wire?
2. When a wire is looped into a coil with many turns, what happens when a current flows through the wire?
3. What does an electromagnet consist of?
4. Apart from increasing the current in the coil, what other alteration can be made to an electromagnet to increase its magnetic field?

DON'T FORGET

There are many uses of magnetism and electromagnetism.

THINGS TO DO AND THINK ABOUT

1. Find out which metals are attracted to magnets.
2. Investigate and draw the the magnetic field line patterns for:
   (i) two magnets with north to north poles facing;
   (ii) two magnets with south to north poles facing;
   (iii) a horseshoe magnet.

PRACTICAL ELECTRICAL AND ELECTRONIC CIRCUITS 1

MEASURING CURRENT, VOLTAGE AND RESISTANCE IN CIRCUITS

Measuring current using an ammeter

Electrons are charged particles that can move through conductors.

To measure the current in a component in an electrical circuit, an **ammeter** is placed in series with the component.

Current is measured in amperes (A).

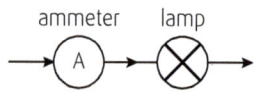

electrons

When electrons move through a conductor such as a copper wire, an electric current is produced.

ammeter   lamp

Using an ammeter to measure the current in a lamp.

Measuring voltage using a voltmeter

The **voltage** or **potential difference** is a measure of the energy required to move a charge through a component (e.g. a lamp).

A **voltmeter** is placed in parallel with the component to measure the voltage **across** it.

Voltage or potential difference is measured in volts (V).

lamp

voltmeter

Using a voltmeter to measure the voltage **across** a lamp.

EXAMPLE

A student is investigating current and voltage using the following equipment: an ammeter; a voltmeter; a 12 V d.c. supply; a lamp; and connecting leads.

Complete a circuit diagram to show how this equipment can be used to measure the current through, and the voltage across, the lamp.

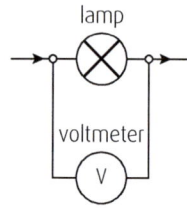

12 V d.c.

Measuring resistance using an ohmmeter

An **ohmmeter** can be used to measure the resistance of any component, such as a resistor or lamp. To measure the resistance, disconnect the component from its circuit and connect it to the ohmmeter as shown in the diagram.

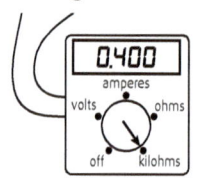

A multimeter can measure current, voltage or resistance.

A **multimeter** can be used to measure current, voltage or resistance by adjusting a selector switch.

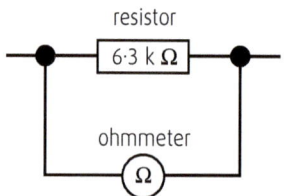

resistor

6·3 kΩ

ohmmeter

Using an ohmmeter to measure the resistance of a lamp.

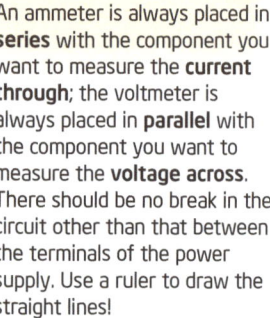

DON'T FORGET

An ammeter is always placed in **series** with the component you want to measure the **current through**; the voltmeter is always placed in **parallel** with the component you want to measure the **voltage across**. There should be no break in the circuit other than that between the terminals of the power supply. Use a ruler to draw the straight lines!

⚙ **TASK**   Quick questions 1

1. Which charged particles move through a wire to cause a current in it?

2. Describe how to use an ammeter to measure the current in a lamp.

3. What is voltage a measure of?

4. Describe how to use a voltmeter to measure the voltage across a lamp.

PRACTICAL ELECTRONIC CIRCUITS AND SYSTEMS

Electrical circuits are drawn using standard symbols. Some of the most common standard electrical and electronic components and symbols used in physics are shown in the table.

Component	Function	Symbol
Cell	Provides the electrical energy to make charge move through a circuit	
Battery	Provides the electrical energy to make charge move through a circuit	
Solar cell (photovoltaic cell)	Renewable energy source; can be used to provide electrical energy to recharge batteries for electronic devices, e.g. vehicle speed warning signs	
Lamp	Used as indicator or for lighting	
Switch	Used to stop or allow current to move in a circuit	
Resistor	Used to control the size of the current in a circuit	
Variable resistor	Used to vary the size of the current in a circuit	
Voltmeter	Measures the voltage **across** a component in a circuit	
Ammeter	Measures the current **through** a component in a circuit	
Fuse	The fuse melts and breaks the circuit if the current in the circuit increases above the value of the fuse	
Diode	Used in electronic circuits	
Light-emitting diode (LED)	Used as an indicator and for low-energy lighting, e.g. car brake lights and sidelights	
Capacitor	Used in electronic amplifiers and timing circuits	
Light-dependent resistor (LDR)	Used in circuits to detect and control light levels	
Thermistor	Used in circuits to detect and control temperature	

Component	Function	Symbol
Motor	The motor speed depends on the size of the supply voltage; its direction can be reversed	
Loudspeaker	Converts output from amplifiers into sound	
Microphone	Converts sound energy into electrical signals for amplifiers	
Relay	When current is present in the relay coil, a switch in a separate circuit is closed	
Transistor	Used in electronic switching circuits and amplifiers. Transistor switches on when its input voltage reaches a fixed level	bipolar transistor / MOSFET (metal-oxide semiconductor field effect transistor)

TASK   Quick questions 2

Look at the table then answer these questions.

1. Draw a circuit diagram with a battery, an LED, a variable resistor and an ammeter connected in series.

2. Draw a circuit diagram of a solar cell connected to a lamp, with a voltmeter connected across the lamp.

3. Draw the symbol for a fuse.

4. What is a transistor?

THINGS TO DO AND THINK ABOUT

When measuring current, voltage and resistance using a multimeter, it is important to adjust the multimeter to the appropriate measurement. For the multimeters used in your laboratory, find out how to select the appropriate settings to measure current, voltage and resistance and how to make sure that the correct range is selected for the measurement.

PRACTICAL ELECTRICAL AND ELECTRONIC CIRCUITS 2

PRACTICAL ELECTRICAL AND ELECTRONIC CIRCUITS AND SYSTEMS

Many electrical systems use electronic components in circuits that monitor conditions such as temperature.

Light-dependent resistors change their resistance when the level of light falling on them is varied.

They can be used as a sensor to determine whether it has become dark enough for street lights to be switched on.

A thermistor's resistance changes as its temperature changes, so it can be used as a temperature sensor.

A capacitor is used in timing circuits and also in devices called accelerometers, which can measure the size of a force to determine whether an object is upright or upside down, or whether the object is accelerating.

These sensors can provide instant information about the surroundings. This information can be used in a variety of ways – for example, in weather forecasting or to provide ice warnings to drivers.

The information can also be used as an input to a control circuit that can alter the temperature/light levels.

The **thermistor** is part of a sensing circuit that produces a **change in voltage** when the temperature changes.

The change in voltage can be used to switch a heater on or off depending on whether the temperature is too low or too high for the requirements.

Electronic components such as light-dependent resistors and thermistors are used as sensors. These components can be used in many electrical circuits to produce useful operations.

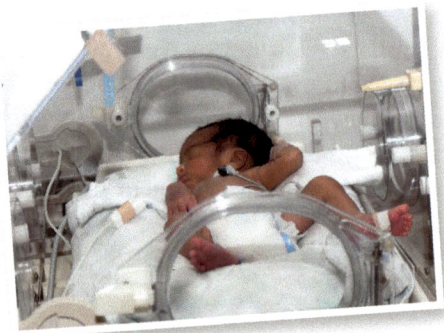

Thermistors are used as sensors in systems that monitor and control the temperature in incubators for newborn babies.

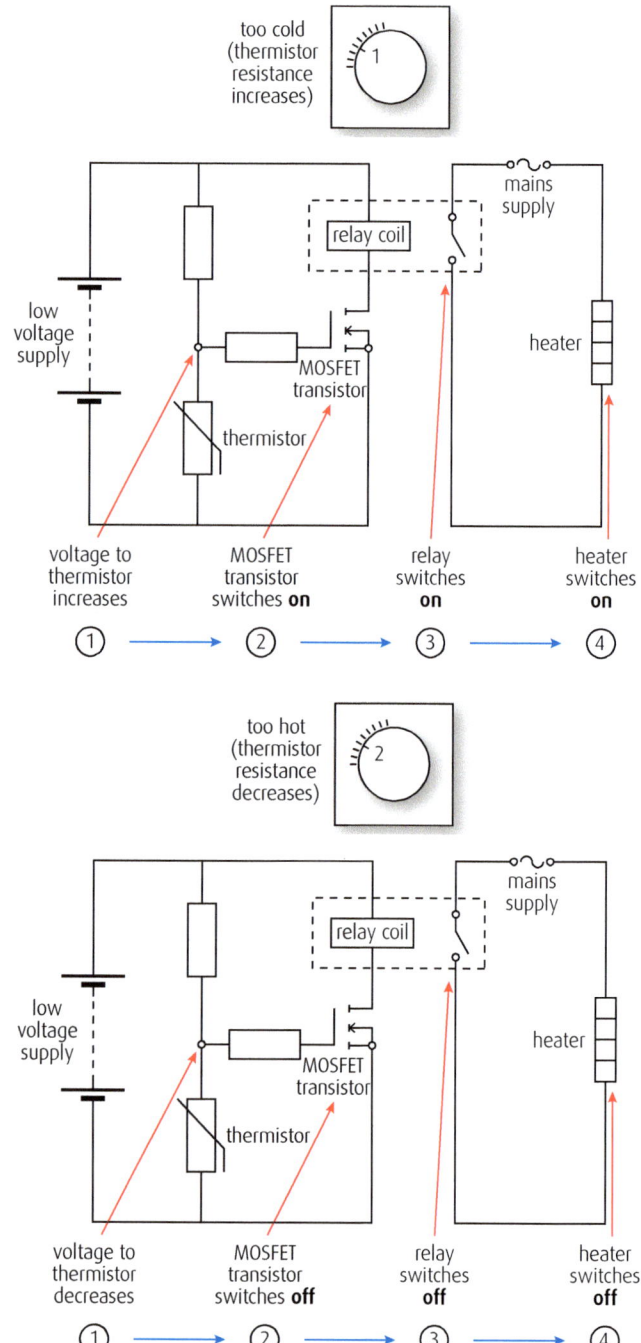

Light Dependent Resistor

Thermistor

Capacitor

switches

TASK   Quick questions 1

1. Which component's resistance changes when the temperature changes?

2. Which component's resistance changes when the light level changes?

3. Which component is used in timing circuits?

PRACTICAL APPLICATIONS OF SERIES AND PARALLEL CIRCUITS

Series circuits

Many common domestic appliances are used in series circuits. An electrical circuit to operate a kettle is a series circuit. Each switch is connected in series.

A central heating control circuit. The electronic control for switching on a central heating boiler has switches connected in series.

The switches are used to:
- disconnect the boiler control from the mains;
- switch off the boiler if the room temperature is hotter than the required temperature;
- switch off the boiler control if the time is outwith the time of day when the heating is set to operate.

If any one switch turns off, then the boiler will be switched off by the control circuit until it is switched on again.

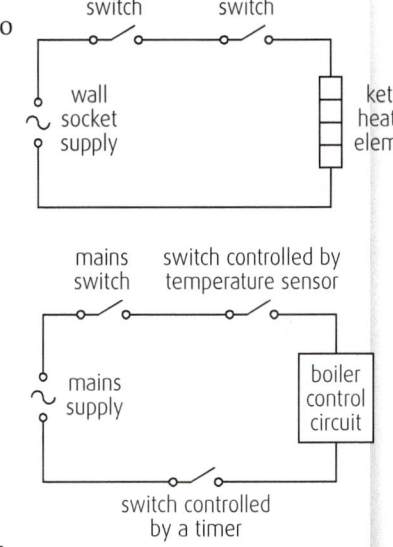

Central heating control circuit.

Parallel circuits

Some household appliances have several switches connected in parallel inside the appliance to control various features.

For example, some hairdryers have one switch simply to switch the motor on to blow cold air and other switches to blow air at different temperatures.

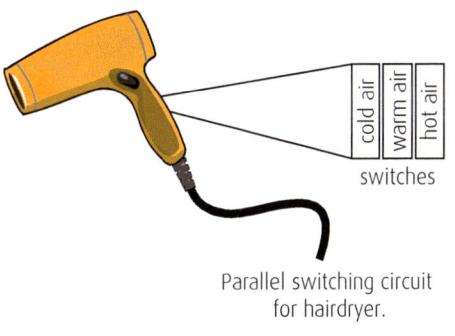

Parallel switching circuit for hairdryer.

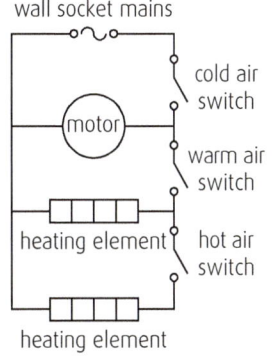

TASK   Quick questions 2

1. Name an application in the home that uses **two** switches in series.

2. Name an application in the home that uses switches connected in parallel.

3. In domestic wiring, what is the advantage of connecting all the sockets in parallel?

THINGS TO DO AND THINK ABOUT

Investigate the properties of a light-dependent resistor by carrying out an experiment to determine its resistance at different light levels. Use a light meter, if available, or simply carry out the experiment under different lighting conditions.

DON'T FORGET

The mains voltage in the UK is 230 V and is applied to all electrical devices in the home.

PRACTICAL ELECTRICAL AND ELECTRONIC CIRCUITS 3

LOGIC CIRCUITS

Logic gates are digital components found in the microchips used in many electrical devices, including computers and calculators.

There are many types of logic gates, including:
- **NOT gate** (also called an **inverter**)
- **OR gate**
- **AND gate**

The NOT gate has only one input, but the other logic gates can have many inputs. You will only need to work with NOT gates and with AND gates and OR gates that have two inputs.

The circuit symbols for the logic gates are shown below:

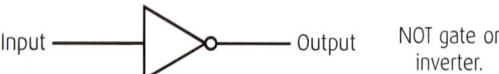

The input and output labels are not required as part of the symbol.

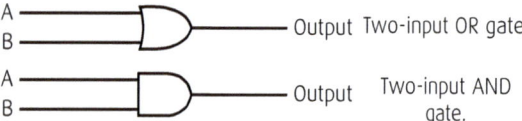

Truth tables can be used to show the outputs for each possible combination of input:

- **1** is used to represent **high** or **on**
- **0** is used to represent **low** or **off**.

NOT gate

Input	Output
0	1
1	0

An easy way to remember how a NOT gate works is that the output is **NOT** the same as the input.

OR gate

Input		Output
A	B	
0	0	0
0	1	1
1	0	1
1	1	1

For an OR gate, the output is high (1) if one input, **OR** the other input, **OR** both inputs are high (1).

AND gate

Input		Output
A	B	
0	0	0
0	1	0
1	0	0
1	1	1

For an AND gate, the output is only high (1) if input A **AND** input B are both high (1).

Circuits containing logic gates are used in a variety of everyday electrical and electronic systems.

Laptops, mobile phones and domestic appliances (e.g. washing machines, kettles and coffee-makers) all use **logic circuits** to control functions or to process data in a particular way.

Logic circuits can be used to represent some common circuits.

> **EXAMPLE**
>
> A central heating control circuit. There is a mains switch to connect with the power supply and there is a timing switch that the householder can set to switch the heating on when it is needed.

The mains switch **AND** the timing switch must be on to allow the heating to operate.

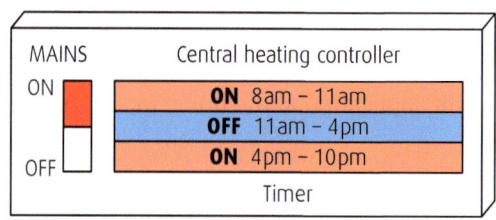

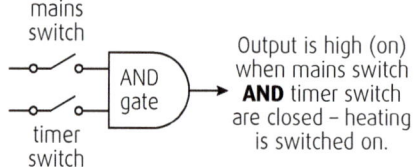

Logic circuit for central heating control.

EXAMPLE

When either the driver door **OR** the passenger door of a car is opened, the interior courtesy lamp lights.

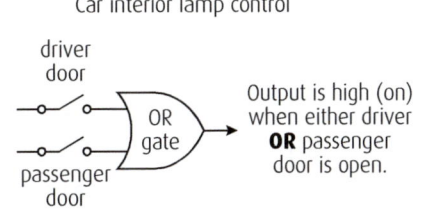

Car interior lamp control

driver
door

OR
gate

Output is high (on)
when either driver
OR passenger
door is open.

passenger
door

Control of car courtesy light is by an OR logic gate.

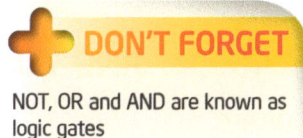

DON'T FORGET

NOT, OR and AND are known as logic gates

Modern appliances are controlled by logic circuits.

These logic circuits start processses happening and control the order of the different functions needed to complete the task.

For example, a microwave oven will heat up food for a specific time. The electronic logic circuit inside the oven will control this process.

Several stages are involved in the process:
- the user inputs the time for cooking;
- the oven will not start until the start button is pressed AND the oven door is closed;
- a timing circuit is started;
- the heater is switched off after the correct time OR if the door is opened.

A microwave oven is controlled by a logic circuit.

TASK   Quick questions

1. In a bank, for security, safety switches can be pressed by any operator if there is any kind of threat. What type of logic gate would be suitable for use in the circuit?

2. In a factory, a dangerous cutting machine has a safety guard. The machine should not operate when the guard is not in position. When the guard is shut, a safety switch is closed. What type of logic gate would be suitable for the circuit so that the machine operates safely when the operator switches it on?

THINGS TO DO AND THINK ABOUT

Think about some everyday chores that are carried out automatically by machines in your house. What information is required by the machine to complete the task? Write down a list of actions that are automatically completed by the machine. For example, think about what happens in a washing machine from when the start button is pressed until the machine switches off. Write down a list of which processes are controlled by the electronic circuits.

PRACTICAL ELECTRICAL AND ELECTRONIC CIRCUITS 4

CURRENT AND VOLTAGE RELATIONSHIPS IN A SERIES CIRCUIT

Current

In a series circuit, the same value of current is measured at all positions in the circuit.

In this circuit, the value of the current measured by all four ammeters is the same.

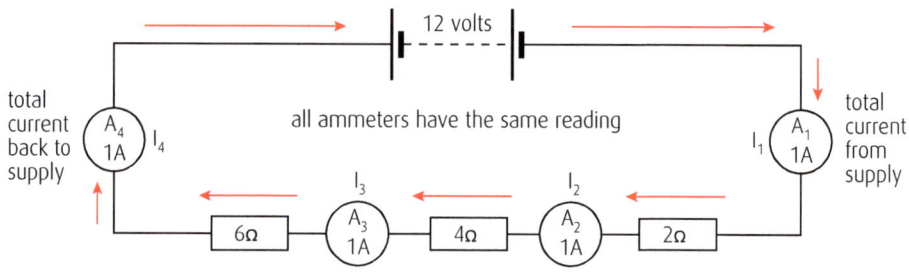

In a series circuit, $I_1 = I_2 = I_3 = I_4 \ldots$

State the current readings on A2, A3 and A4 in the following circuit:

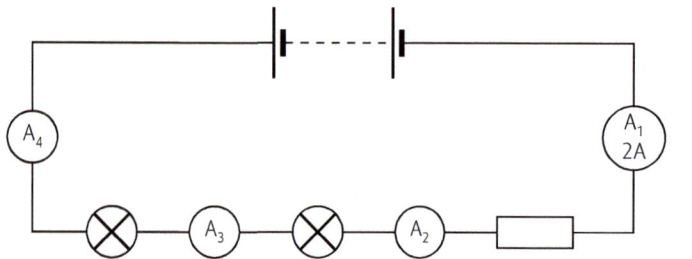

Voltage

In a series circuit, the supply voltage divides across each component.

The voltages across each of the components adds up to the supply voltage.

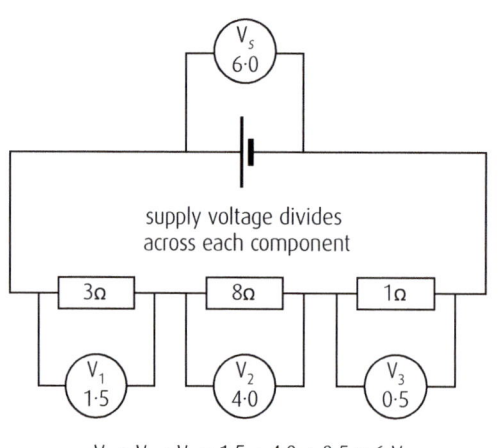

supply voltage divides across each component

$V_1 + V_2 + V_3 = 1\cdot5 + 4\cdot0 + 0\cdot5 = 6$ V
In a series circuit, $V_S = V_1 + V_2 + V_3 \ldots$

This is a series circuit. This means that the current has the same value at all points in the circuit.

So $A_1 = A_2 = A_3 = A_4 = 2$ A

DON'T FORGET

In series circuits, the voltages across each of the components add up to the supply voltage.

JUST A WEE NOTE

The greatest value of resistance receives the greatest share of the voltage

State the voltage readings on V_2, V_3, V_4, and V_5.

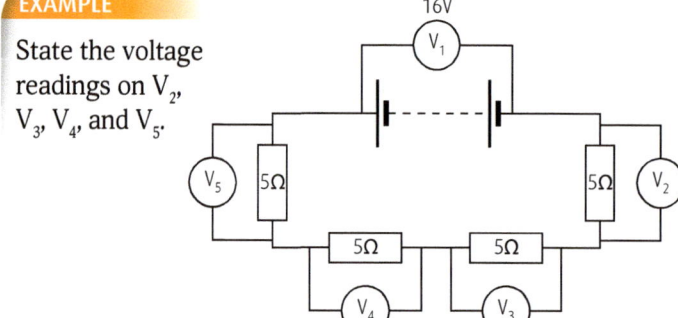

The supply voltage divides across each resistor. As each resistor has the same value, the supply voltage will divide equally between them.

So $V_2 = V_3 = V_4 = V_5 = 4$ V

A series circuit containing lamps can be used as a decoration on a Christmas tree. Each of the lamps operates at 5 V. The circuit is connected to the 230 V mains supply.

QUESTION

Calculate how many lamps are required in this circuit to allow them to operate at their suggested operating voltage.

ANSWER

The lamps are identical and so have the same value of resistance.

This means that the supply voltage will divide equally across all lamps.

Mains voltage = 230 V; voltage of one lamp = 5 V.

$$\text{Number of lamps} = \frac{230}{5} = 46$$

Therefore 46 lamps are required to operate each lamp at 5 V.

QUESTION

State a disadvantage of connecting the lamps in this way.

ANSWER

In a series circuit, if one lamp breaks, then the circuit is incomplete and the other lamps do not light.

CURRENT AND VOLTAGE RELATIONSHIPS IN A PARALLEL CIRCUIT

Current

In a parallel circuit, the supply current **splits up** through each component when it reaches a branch in the circuit. In this circuit, the total 9 A supply current splits through each branch of the circuit.

Voltage

In a parallel circuit, the voltage across components connected in parallel **remains the same** as the supply voltage.

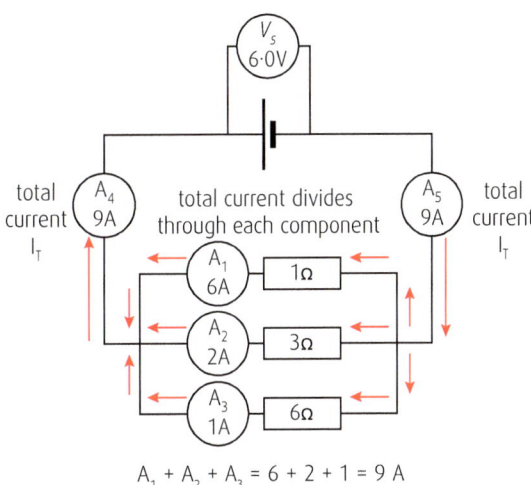

$A_1 + A_2 + A_3 = 6 + 2 + 1 = 9\ A$
In a parallel circuit, $I_T = I_1 + I_2 + I_3 \ldots$

The supply voltage of 6 V is measured across each resistor connected in parallel.

ONLINE

To see the differences between series and parallel circuits in practice, watch the video at http://www.brightredbooks.net/subjects/n5physics/c02_04.

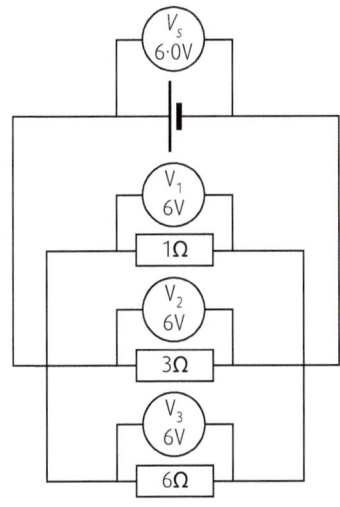

In a parallel circuit,
$V_s = V_1 = V_2 = V_3 \ldots$

THINGS TO DO AND THINK ABOUT

1. A battery, resistor and lamp are connected in a series circuit. There is a current of 2 A in the resistor. What is the current in the lamp?

2. Three identical lamps are connected in series to a 30 V supply. What is the voltage across each lamp?

RESISTANCE AND OHM'S LAW

RESISTORS

Resistors can be used to control the size of the current in a circuit. Increasing the resistance in a circuit reduces the current.

The resistance of a material (or conductor) depends on several factors.

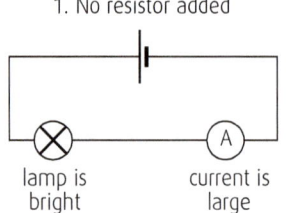

1. No resistor added

lamp is bright — current is large

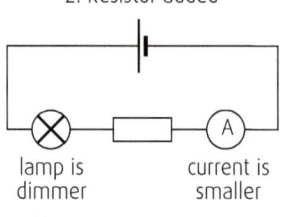

2. Resistor added

lamp is dimmer — current is smaller

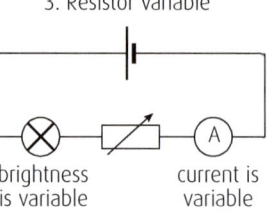

3. Resistor variable

brightness is variable — current is variable

Adding resistance to a circuit.

Type of material

Different conductors have different resistance properties. For example, wires made from the metal tungsten have a greater resistance than copper wires of the same size.

Copper wire has a very low resistance and so is often used in wiring circuits.

Length of material

The resistance of a wire depends on its length – the **longer** the wire, the **greater** its **resistance**.

Thickness of material

The **greater** the **cross-sectional area** of a wire, the **smaller** its **resistance**.

Thicker wire is used in circuits where the current is likely to be very large. If there is a large current in a thin conductor, then this could cause the wire to overheat because of its higher resistance.

For domestic wiring, the copper cable required when installing an electric cooker has a much greater cross-sectional area than the cable used for lighting circuits. This is because the current in the cooker wire can be much greater than that in the lighting circuit.

Other factors affecting resistance

For most components, the resistance remains approximately constant as the current is changed, providing that the temperature remains the same. There are a number of components that can, and do, vary their resistance, such as **variable resistors**, **light-dependent resistors** and **thermistors**. Variable resistors have lots of uses, including the volume control on a sound system and the brightness and contrast controls on some computer monitors.

> **DON'T FORGET**
>
> Resistance is measured in ohms (Ω).

`0·19` Ω ohmmeter

`0·63` Ω ohmmeter

copper wire
less resistance

tungsten wire
more resistance

copper wire

longer copper wire has **more** resistance

Longer wires have a higher resistance.

thin wire

thicker wire has **less** resistance

Thicker wires have less resistance.

🔧 TASK Quick questions

1. What happens to the current in a circuit if the total resistance is increased?
2. State three factors that the resistance of a conductor depends on.
3. What happens to the resistance of a wire when its length is increased?
4. How does the resistance of a wire change when its cross-sectional area is increased?
5. What can happen to a thin wire if there is too much current in it?
6. Why is copper wire commonly used in domestic wiring circuits?

OHM'S LAW: CURRENT (I), VOLTAGE (V) AND RESISTANCE (R) IN CIRCUITS

Resistance is the property of any conductor (such as a lamp, resistor, or even a wire) to oppose current. When the total resistance in a circuit is increased, the current is reduced.

Resistance can be calculated using the relationship: $R = \dfrac{V}{I}$

where R is the resistance in ohms (Ω), V is the voltage (potential difference) in volts (V) and I is the current in amperes (A).

This is **Ohm's law**.

This relationship can be used to calculate the current, voltage or resistance in a circuit.

EXAMPLE

Calculate the resistance of the resistor in circuit 1.

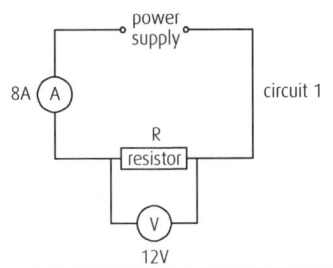

ANSWER

$R = \dfrac{V}{I} = \dfrac{12}{8} = 1{\cdot}5\ \Omega$

EXAMPLE

Calculate the current in the resistor in circuit 2.

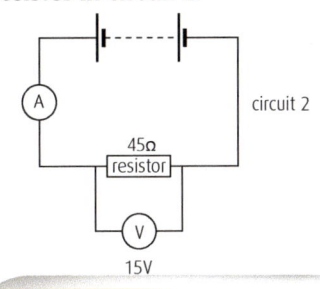

ANSWER

$I = \dfrac{V}{R} = \dfrac{15}{45} = 0{\cdot}3\ \text{A}$

EXAMPLE

Calculate the voltage across the resistor in circuit 3.

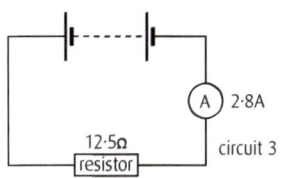

ANSWER

$V = I \times R = 2{\cdot}8 \times 12{\cdot}5 = 35\ \text{V}$

EXAMPLE

The following graph shows the voltage and current readings taken from a circuit containing a variable power supply connected to a fixed resistor. After each increase in the power supply voltage, the current and voltage readings were recorded and plotted on the graph.

(a) Calculate the value of the resistance when the voltage is:
   (i) 1·2 V (ii) 3·0 V (iii) 6·0 V
(b) Comment on the answers to
   (a) (i), (ii) and (iii).

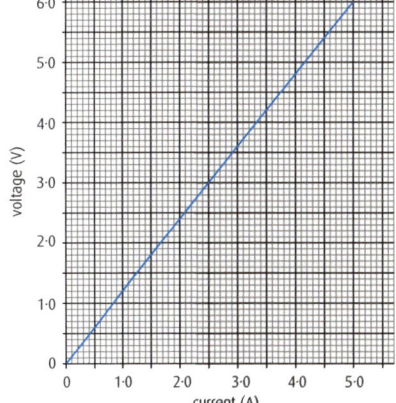

ANSWER

(a) (i) When $V = 1{\cdot}2$ V, $I = 1{\cdot}0$ A

$R = \dfrac{V}{I} = \dfrac{1{\cdot}2}{1{\cdot}0} = 1{\cdot}2\ \Omega$

(iii) When $V = 3{\cdot}0$ V, $I = 2{\cdot}5$ A

$R = \dfrac{V}{I} = \dfrac{3{\cdot}0}{2{\cdot}5} = 1{\cdot}2\ \Omega$

(iii) When $V = 6{\cdot}0$ V, $I = 5{\cdot}0$ A

$R = \dfrac{V}{I} = \dfrac{6{\cdot}0}{5{\cdot}0} = 1{\cdot}2\ \Omega$

(b) The resistance remains constant as the current and voltage increase.

THINGS TO DO AND THINK ABOUT

1. State the relationship for Ohm's law.
2. A resistor has a voltage of 6·1 V across it and a current of 2·1 A in it. Calculate the resistance of the resistor.
3. Calculate the current in a 5 Ω resistor when the voltage across it is 6·5 V.
4. Calculate the voltage across a 15 Ω lamp when the current in it is 0·3 A.

GAS LAWS AND THE KINETIC MODEL

This section is about how gases behave. Gases can exist in containers, such as metal cylinders, aerosol spray cans and bicycle inner tubes.

When studying the behaviour of gases, the temperature (T), pressure (p) and volume (V) are important quantities used to explain and determine how gases behave.

These quantities can be easily measured using thermometers (temperature), pressure gauges (pressure) and syringes (volume).

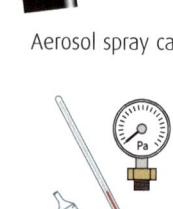

Gas cylinder

Aerosol spray can

Bicycle inner tube

Thermometers, syringes and pressure gauges are used to measure the properties of gases.

PRESSURE

Pressure occurs when a force is applied to a surface. Whenever a force is applied to an object, a pressure is applied to that object. For example, a force can be caused by a pencil point writing on paper, or by a molecule of gas colliding with the wall of its container.

This person is standing in shoes with a very small area of contact with the ground.

A small weight on a small area can cause damage to the floor.

A person standing on the ground exerts a **force** on the **area** of the ground covered by their feet. This force is the person's weight.

A building exerts a pressure on the ground it stands on.

Force is measured in newtons (N).

Area is measured in square metres (m²).

Pressure is measured in Pascals (Pa) or newtons per square metre (N m⁻²).

Pressure, p, is calculated by dividing the force, F, by the area, A:

$$p = \frac{F}{A}$$

The pressure depends on the size of the force and the area:

$$p = \frac{\text{force}}{\text{large area}} = \text{small pressure}$$

$$p = \frac{\text{force}}{\text{small area}} = \text{large pressure}$$

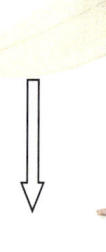

The person's weight is the force which acts through the feet onto the ground.

The weight of the building is the force which acts through the foundations onto the ground.

The tyres of the tractor have a very large area of contact with the ground.

A heavy tractor does not sink into the ground.

EXAMPLE

A person whose weight is 637 N is standing on both feet. Each foot has an area of 0·02 m².

Calculate the pressure exerted on the floor.

ANSWER

$$p = \frac{F}{A} = \frac{637}{2 \times 0·02} = 15\,925 \text{ Pa}$$

When an inflated balloon is moved from a cold spot to a hotter spot, its volume increases.

EXAMPLE

A four-wheeled car is parked. Each tyre has an area of 0·035 m² in contact with the ground The weight of the car is 9800 N.

Calculate the pressure exerted on the ground by the car.

ANSWER

$$p = \frac{F}{A} = \frac{9800}{4 \times 0·035} = 70\,000 \text{ Pa}$$

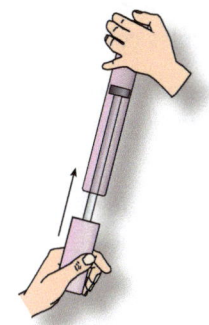

There are many common examples that show the connection between the temperature, pressure and volume of gases.

The following facts help to explain the behaviour of gases:

- a gas consists of billions of moving molecules or atoms;
- gas molecules are constantly moving at different speeds;
- the distance between the molecules is much greater than their size;
- the molecules move in random directions (no fixed direction), and in straight lines, until they collide with other molecules or rebound against the container holding them;
- the temperature of the gas is a measure of the average kinetic energy of the gas molecules – when a gas is heated, this kinetic energy increases;
- because of the huge number of gas molecules in even a small container, there are approximately equal numbers of molecules always moving in every direction.

Making the volume of air trapped inside a pump smaller by pressing the plunger in increases the pressure of the air inside the pump.

If a gas is trapped in a sealed container, then the number of gas molecules cannot increase or decrease, so the number is fixed. This means that the **mass** of the gas trapped inside is also fixed.

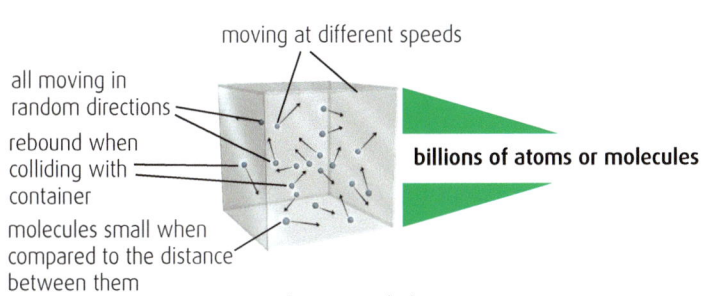

moving at different speeds

all moving in random directions

rebound when colliding with container

molecules small when compared to the distance between them

billions of atoms or molecules

Gas particles in a sealed container.

Deep-sea divers have to return to the surface slowly until the air pressure inside their body equals the surface air pressure. This prevents the gases inside their body expanding too quickly and causing 'the bends'.

THINGS TO DO AND THINK ABOUT

1. Explain in terms of pressure why a person wearing snow-shoes does not sink into the snow.

2. An elephant whose weight is 58 800 N is standing on all four feet. Each foot has an area of 0·21 m². Calculate the pressure exerted by the elephant on the ground.

DON'T FORGET

Gas pressure is caused by molecules colliding with the container walls.

GAS PRESSURE

The pressure exerted by a gas is caused by the millions of moving gas molecules that collide with the inside walls of the gas container. Each time a molecule collides with the wall, the wall receives a tiny force.

The combination of millions of these tiny forces on the container walls causes the gas pressure. The pressure is a result of the average force produced by millions of molecules colliding with the walls of the container.

If the **gas molecules** are **heated** and their temperature rises, they **gain kinetic energy**.

ONLINE

Watch the simulation of gas particles at http://www. brightredbooks.net/subjects/ n5physics/c02_12

This **increased speed** and **energy** also means that there are **more collisions** and they collide with **more force** so the pressure increases.

Pressure is the average force exerted on an area of 1 m².

Pressure is measures in units called Pascals (Pa) or newtons per square metre (N m⁻²).

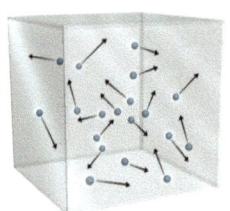

Gas particles in a solid container.

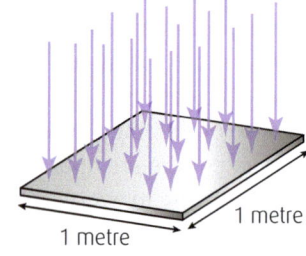

molecules colliding with balloon wall

molecules moving randomly at different speeds

Gas molecules moving inside an inflated balloon.

AIR PRESSURE

The air in the atmosphere extends about 120 km above us. The pressure exerted by gas molecules in the air is known as the atmospheric pressure. Millions of gas particles are continually colliding with the Earth's surface (and us!), each one exerting a tiny force. The result of all of these collisions is to produce atmospheric (or air) pressure.

Normal atmospheric pressure is 101 000 Pa ($1{\cdot}01 \times 10^5$ Pa). This is equivalent to a force of 101 000 N acting on 1 m² of surface.

1 metre

1 metre

Air particles colliding with a unit surface area.

PRESSURE, VOLUME AND TEMPERATURE

When gases are trapped inside a container, what happens to these gas molecules can be described by the kinetic theory of gases.

Changing the volume of a gas

When the **volume** of a gas is **reduced**, without changing the temperature, the molecules do not have to travel so far to collide with the container walls and so the **pressure increases**.

If the **volume** of the gas is **increased**, without changing the temperature, the gas **pressure decreases**.

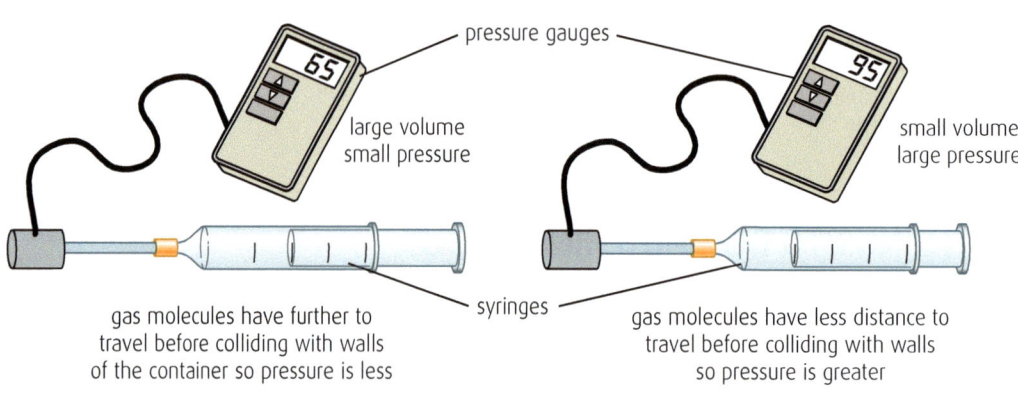

pressure gauges

large volume small pressure

small volume large pressure

syringes

gas molecules have further to travel before colliding with walls of the container so pressure is less

gas molecules have less distance to travel before colliding with walls so pressure is greater

Effect of volume on gas pressure.

Changing the temperature of a gas

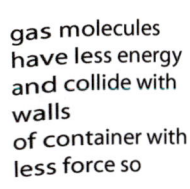

gas molecules have less energy and collide with walls of container with less force so

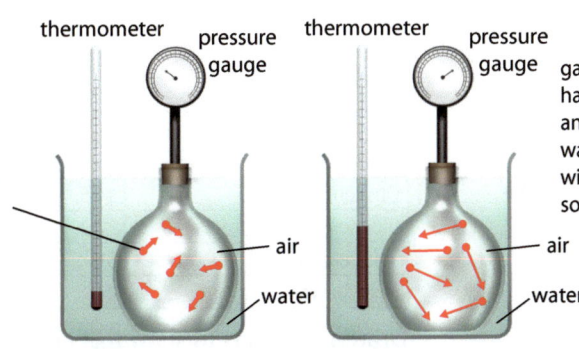

thermometer pressure gauge   thermometer pressure gauge

gas molecules have more energy and collide with walls of container with greater force so pressure

air

water

air

water

low temperature   high temperature

Effect of temperature on gas pressure.

DON'T FORGET

The pressure, temperature and volume of a gas depend on the behaviour of the gas molecules.

When the **temperature** of a gas is **increased**, if the volume of the container does not change, then the **pressure increases**. This is because the molecules have greater kinetic energy (and speed) and so collide with the walls with more force and more often.

If the volume of the gas is able to change, because the size of the container can increase, then when the temperature of the gas increases, the volume will increase until the pressure of the gas is the same as the pressure outside the container.

EXAMPLE Pressure cooker

A pressure cooker contains water that is converted into a gas (steam) by heating. As the temperature of the steam increases, the pressure of the steam also increases. To prevent the lid from blowing off as a result of the increased pressure, there is a pressure release valve that lets some of the steam out, which reduces the pressure.

As the pressure cooker is heated, the pressure of the gas inside also increases. The pressure cooker has a release valve to let hot gas out to reduce the pressure.

EXAMPLE Weather balloons

Weather balloons are used to transport instruments to a great height (altitude) in the atmosphere. The balloons send information about the weather back to a ground station.

As the balloon rises, the pressure of the atmosphere becomes less and the volume of the balloon increases.

Weather balloon at launch.   Weather balloon at high altitude.

THINGS TO DO AND THINK ABOUT

1. What happens to gas molecules when they are heated in a container?
2. What happens when a gas molecule collides with the walls of its container?
3. What happens to the volume of a gas, at constant temperature, if its pressure is increased?
4. What happens to the pressure of a gas, with a fixed volume, if its temperature is decreased?
5. What happens to the volume of a gas, if its pressure is not fixed, if its temperature is increased?

EXTENDED QUESTIONS 1

GENERATION OF ELECTRICITY

1. (a) In a coal-fired power station, state the energy transformations when:
     (i) coal is burned; and
     (ii) the turbines turn the generator.
   (b) Why is steam required?

2. In coal-fired power stations, there are many waste products, such as coal ashes, sulfur dioxide and carbon dioxide.

   (a) What is the main waste product from nuclear power stations?
   (b) What is the main problem associated with this waste?

3. (a) In a hydroelectric power station, state the energy transformations when:
     (i) water from the upper reservoir reaches ground level; and
     (ii) this water arrives at the turbines and generator.

4. Longannet coal-fired power station produces up to 2·3 GW of electrical power; Black Law wind farm produces up to 0·124 GW of electric power.

   Calculate the number of similar-sized wind farms that would be required to replace the Longannet power station.

5. What is the difference between a hydroelectric power station and a pumped storage hydroelectric power station?

6. Explain the process of how heat energy is produced in nuclear power stations. In your explanation, mention the words: fission, uranium nucleus, chain reaction, neutron and heat energy.

7. In a coal-fired power station, 1 kg of coal produces $3·0 \times 10^7$ J of heat energy.
   In a nuclear power station, 1 kg of uranium produces $4·5 \times 10^{12}$ J of heat energy.

   Calculate how many kilograms of coal are required to produce the same amount of heat energy as 1 kg of uranium.

8. The table shows the amount of electricity generated in PJ (petajoules) by different sources of renewable energy in the UK during the summer of 2014.

Energy source	Electricity generated (PJ)
Offshore wind	7·2
Onshore wind	11·5
Bioenergy	20·2
Solar energy	3·6
Hydropower	4·3

   Draw a bar chart to show this information.

9. Copy and complete the table to show the **advantages** and **disadvantages** of the various sources of electrical energy.

Source of energy	Advantages	Disadvantages
Wind		
Solar		
Coal		
Nuclear		
Hydro		

10. The diagram shows equipment that can be used to generate electricity.

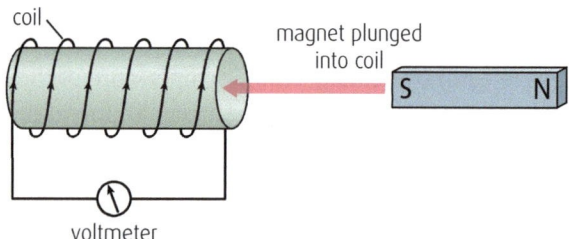

   (a) Describe how a voltage is produced in the generation of electricity. In your description, use the words coil of wire, magnet, movement, iron core, voltage.
   (b) How can the size of the voltage be increased?

11. The microgeneration of electricity is seen a useful addition to help supply some of our energy needs in the future.

   Give an example of microgeneration and describe where it is used.

12. The National Grid is responsible for the distribution of electrical energy from power stations.

   (a) What is the main cause of energy loss in transferring this energy across the country in overhead wires?
   (b) How is this energy loss reduced?

13. The three main parts of a transformer are: the secondary coil; the primary coil; and the core. Label these three parts of the transformer on the diagram.

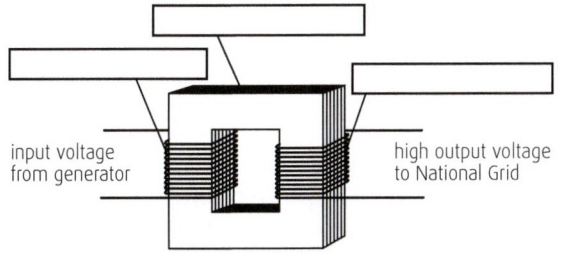

input voltage
from generator

high output voltage
to National Grid

14. The trams used in a city operate at a voltage of 750 V. A transformer at a substation converts the voltage from 11 000 V to the required 750 V.

   There are 88 000 turns on the primary coil of the transformer.

   Calculate the number of turns on the secondary coil.

15. The mains charger for a rechargeable drill contains a transformer with 6440 turns of wire in the primary coil and 504 turns in the secondary coil. The mains voltage is 230 V.

   Calculate the output voltage of the charger.

16. Transformers can be drawn using the symbol in the diagram.

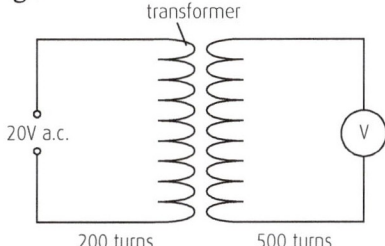

transformer

20V a.c.

V

200 turns   500 turns

   Calculate the reading on the voltmeter.

17. Transformers are used in the National Grid system to distribute electrical power.

   (a) Explain why transformers are used in the National Grid.
   (b) The voltages required at different stages of the National Grid are shown in the table.

Stage	Voltage (V)
Super Grid	400 000
National Grid	132 000
Heavy industry	33 000
Light industry	11 000
Domestic houses	230

   A transformer is used to transfer power from the National Grid to light industry. The primary coil of the transformer has 9000 turns.
   (i) State the voltage input at the primary coil.
   (ii) State the voltage output at the secondary coil.
   (iii) Calculate the number of turns in the secondary coil.

ELECTRICAL POWER

1. Explain what is meant by the 'power rating' of a domestic appliance.

2. The rating plate for a kettle is shown in the following:

   This appliance works for a range of voltages but remember the declared value of the mains in Britain is 230 V

   VAC

   Model: V-015T

   220–240 V - 50 Hz

   1600 W

   The power rat of the applian is shown here

   Serial number:

   01463/33   Made in China

   This symbol is the **Double Insulation** symbo This means that the appliance uses 2-core c with live and neutral wires, and does not re an earth wire.

   (a) How many joules of energy are transferred by the kettle each second?

   (b) The kettle has markings to indicate how much water it has been filled with. Suggest why this could be useful for energy-saving when the kettle is being used.

3. The table shows some facts about domestic appliances.

Appliance	Operating a.c. voltage (V)	Power (W)	Operating frequency (Hz)
Iron	230	1200	50
Kettle	230	2400	50
Lamp	230	40	50
Television	230	75	50

   State which appliance would transfer the most energy when switched on for five minutes.

4. Calculate the amount of energy transferred by a 12 W lamp in one minute.

5. A 900 W toaster is switched on for three minutes. Calculate the energy transferred by the toaster during this time.

6. Explain why the total electrical energy output of a power station is always less that the total input energy of the power station.

7. A power station produces 620 MJ of input heat energy. The generator connected to the turbine produces 570·4 MJ of electrical energy.

   Calculate the efficiency of the power station.

8. In a pumped storage hydroelectric power station, the motor that drives the pump uses 58·8 MJ of electrical energy to pump water back up to the top reservoir.

   The water gains 49·98 MJ of potential energy.

   Calculate the efficiency of this water-pumping process.

9. A microwave oven used 60 750 J of energy to add 44 955 J of useful heat energy to water in a cup.

   Calculate the efficiency of the oven.

EXTENDED QUESTIONS 2

ELECTROMAGNETISM

1. Copy the diagram of the permanent magnet and sketch the magnetic field around it.

2. The following simple circuit for an electromagnet can be used to collect small pins that have been dropped onto the floor.

simple electromagnet

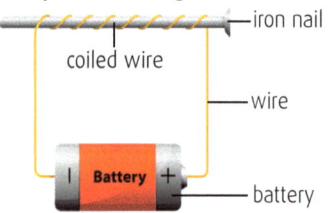

iron nail

coiled wire

wire

battery

(a) Suggest how the strength of the electromagnet could be increased.

(b) State what happens to the pins when the battery is disconnected.

3. Describe two applications of magnetism that require permanent magnets.

PRACTICAL ELECTRICAL AND ELECTRONIC CIRCUITS

1. When there is a current in a wire, what is the name of the charged particles moving through the wire?

2. Copy the following circuit diagram and enter the readings and units on all of the ammeters.

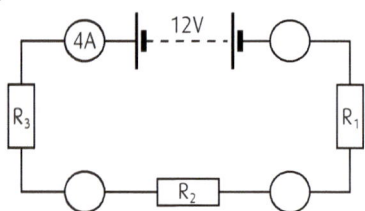

3. In the following circuit, the resistors are identical.

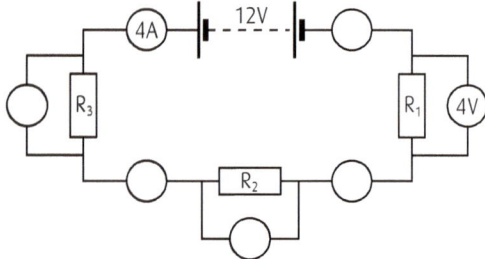

Copy the circuit diagram and enter the readings and units on all of the meters.

4. Copy the following circuit diagram and draw labelled meters to measure the voltage across the resistor and the current in the lamp.

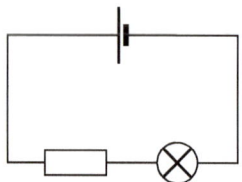

5. Draw the symbol for: (i) a thermistor; (ii) a variable resistor; and (iii) a light-dependent resistor LDR.

6. A light-dependent resistor was used in an electronic circuit to monitor the light levels inside a greenhouse. The light levels were measured in units called 'lux'.

The table shows the resistance of the light-dependent resistor for different light levels.

Light level (lux)	Resistance (Ω)
700	280
600	240
500	200
400	
300	125
200	90

Suggest a value for the resistance of the light-dependent resistor when the light level is 400 lux.

7. Photovoltaic cells (solar cells) installed on the roofs of some houses produce electrical energy. This energy can be used immediately by the householder or can be sold to the National Grid to be used by other people. Information about one type of solar cell is given in the table.

Average annual electrical energy output per m²	5.2×10^8 J m⁻²
Energy contained in one unit of electrical energy	3.6×10^6 J
Cost of one unit of electrical energy	19.4 pence

One house has an area of 18 m² of these solar cells.

(a) Calculate the total electrical energy output available from this house in one year.

(b) The householder sells all of this energy back to the National Grid. Calculate how much the National Grid pays the householder for the electricity generated in this year.

(c) Explain why the total electrical energy generated in the following year might be different.

8. Calculate the voltage across a 550 Ω resistor when the current in it is 0·03 A.

9. Calculate the current in a 48 Ω resistor when it is connected to a 6 V supply.

10. A car headlamp takes a current of 2 A from a 12 V car battery. Calculate the resistance of the lamp.

11. The following graph was produced using readings taken from an experiment to determine the resistance of a resistor.

    (a) Calculate the value of the resistor.
    (b) Determine the voltage reading that would be obtained if the current reading was 11 A.

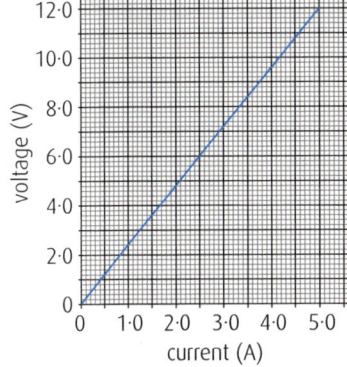

12. A householder wants the doorbell to ring if the switch at the front door or the switch at the back door is pressed.

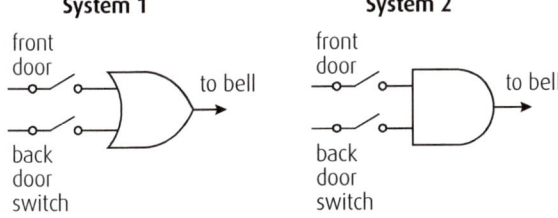

Which of the above logic systems would be suitable for controlling the bell?
Explain your answer.

13. State two electronic devices from the following list that are associated with the **input** of light energy: solar cell; variable resistor; light-dependent resistor; thermistor; light-emitting diode; relay; motor; switch.

14. Electrical energy is transmitted along overhead cables that have a resistance of 0·3 Ω per kilometre.

    One cable is 30 km long and there is a current of 18 A in the cable.

    Calculate the voltage drop across the cable.

15. The circuit below has three resistance wires made from the same material.

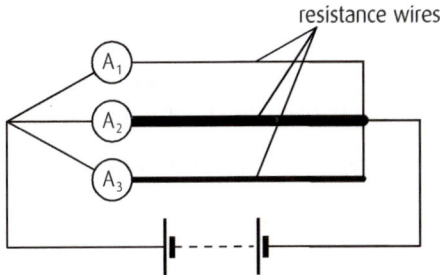

Which ammeter will show the highest reading?

GAS LAWS AND THE KINETIC MODEL

1. A shed containing a gas bottle catches fire.

    (a) State what happens to the gas pressure inside the bottle when it is heated in the fire?
    (b) Explain what happens to the gas particles when they are heated.

2. A cyclist inflates her bicycle tyres to the recommended pressure. The cyclist then stores the bicycle in a garden shed. Next morning, after a very a frosty night, she checks the tyre pressure.

    Explain what has happened to the tyre pressure.

3. A party balloon is inflated in a garden in the morning when the temperature is fairly cold. The balloon is left in the garden during the afternoon when the temperature rises, but the pressure remains constant.

    (a) What happens to the volume of the balloon?
    (b) Explain your answer in terms of the behaviour and movement of the gas molecules.

WAVES AND RADIATION

WAVE CHARACTERISTICS 1

Waves are created by vibrations. They are an important way of describing how energy is transferred from one place to another. For example, light waves transfer energy from light sources, such as the Sun or light bulbs, to our eyes. Sound waves transfer energy from sounds made by voices to our ears. Water waves transfer energy across the sea from the place where the waves were created (usually by winds and tides). Waves that reach the shore can transfer large amounts of potentially damaging energy – strong barriers are built to protect harbours from this damage.

There are two types of wave: **transverse** waves and **longitudinal** waves.

VIDEO LINK

Check out the animation of transverse waves at http://goo.gl/YQa2IS

TRANSVERSE WAVES

A 'slinky spring' can be used to demonstrate transverse waves. One end of the spring is fixed and the other end of the spring is moved sideways.

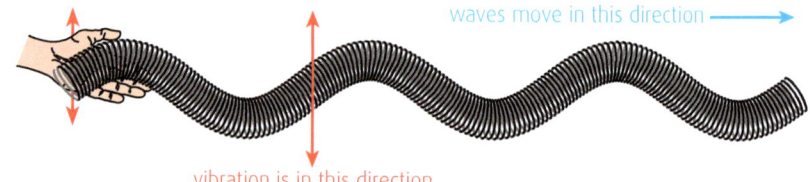

waves move in this direction

vibration is in this direction

Transverse waves in a 'slinky spring'.

The direction in which the waves travel is at right angles to the direction of the movement that produces them. Although the direction of the transverse waves is forward, the actual particles of the spring move (vibrate) at right angles to this direction.

Energy transfer by transverse waves

A duck floating in a pond moves up and down, not sideways, when water waves reach it because the water molecules move up and down as the waves pass.

The electromagnetic spectrum (including light waves) consists of transverse waves.

direction of wave travel

direction of particle movement

direction of particle movement

direction of wave travel

wave moving forward

wave moving forward

Movement of a single particle in a spring for a transverse wave.

VIDEO LINK

Check out the clip on particle motion at http://goo.gl/IMAIQe

LONGITUDINAL WAVES

A 'slinky spring' can also be used to demonstrate longitudinal waves. One end of the spring is fixed and the other end of the spring is pushed back and forth in the direction of the spring. The spring is alternately 'squashed' and 'stretched' in the direction of wave travel. The 'squashed' and 'stretched' parts are known as 'compressions' and 'rarefactions'.

VIDEO LINK

See demonstrations of both types of wave with a 'slinky spring' at http://goo.gl/wCU98z and at http://goo.gl/kARaRv

The direction in which the waves travel is in the same direction as the movement that produces them. The actual particles of the spring move (vibrate) back and forth in the same direction as the longitudinal wave.

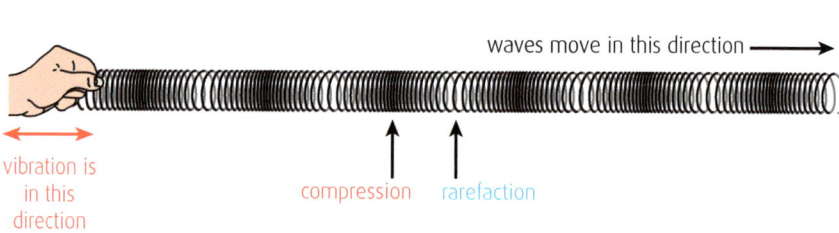

waves move in this direction

vibration is in this direction

compression   rarefaction

Longitudinal waves in a 'slinky spring'.

longitudinal wave   wave direction ⟶

⟵⟶ particle direction

Movement of a single particle in a spring for a longitudinal wave.

Energy transfer by longitudinal waves

Sound waves are longitudinal waves.

During earthquakes, the compression and then stretching of the ground causes longitudinal waves called P waves that travel through both liquids and solids in the interior of the Earth before eventually reaching the Earth's surface.

DESCRIBING WAVES

The diagram shows how the various parts of a wave are usually represented.

Waves are usually represented by drawing a transverse wave. The various parts of the wave are:

- **crest** and **trough** – these are the highest and lowest positions of the waveform;
- **amplitude**, a – this is the distance from the middle to the top or bottom of the wave;
- **wavelength**, λ (lambda) – this is length from one point on the wave until the shape repeats itself.

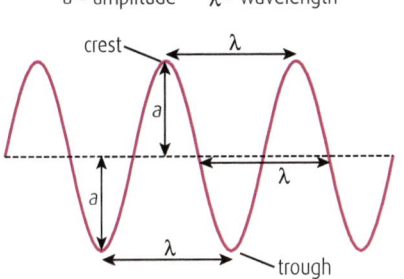

a = amplitude λ = wavelength

EXAMPLE

Determine the amplitude of the following waveform:

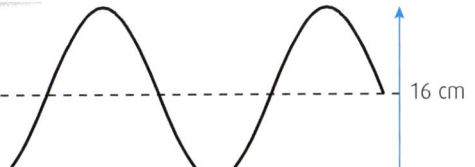

16 cm

ANSWER

$a = \dfrac{16}{2} = 8$ cm

EXAMPLE

Identify the distances that represent one wavelength for the following waveform:

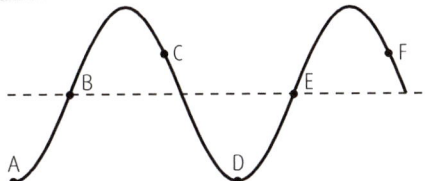

ANSWER

A–D, B–E or C–F

EXAMPLE

Determine the wavelength of the following waveform:

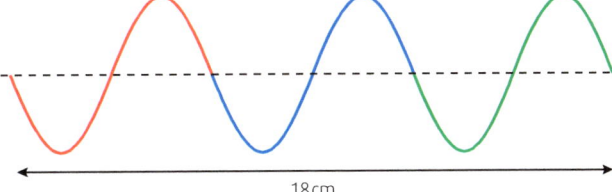

18cm

ANSWER

Three complete waves in 18 cm:

$3\lambda = 18$

$\lambda = \dfrac{18}{3} = 6$ cm

THINGS TO DO AND THINK ABOUT

1. What is transferred from one place to another by waves?
2. Name two types of waves.
3. What is the direction of the vibration of particles relative to the wave direction for transverse waves?
4. What is the direction of the vibration of particles relative to the wave direction for longitudinal waves?
5. Give one example of transverse waves and one example of longitudinal waves.

DON'T FORGET

Transverse waves: particles vibrate at right angles to the wave's direction of travel. Longitudinal waves: particles vibrate in the same direction as the wave's direction of travel.

WAVE CHARACTERISTICS 2

WAVE FREQUENCY

The frequency (f) of a wave is the number of waves produced in a given time period, which is usually one second.

If we say that a wave has a frequency of 5 Hz, then five complete waves are produced every second.

$$\text{Frequency} = \frac{\text{number of waves }(N)}{\text{time taken }(t)} \left(f = \frac{N}{t} \right)$$

Frequency is measured in hertz (Hz), where 1 Hz is one wave per second.

A wave machine in a swimming pool produces 120 waves in 40 seconds.

What is the frequency of the waves?

$$f = \frac{N}{t} = \frac{120}{40} = 3 \text{ Hz}$$

An oscilloscope displays an electronic signal as transverse waves.

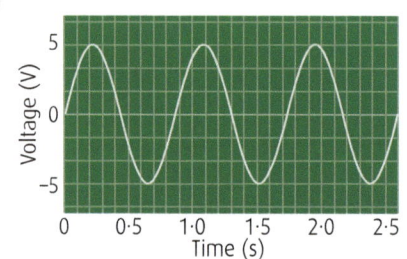

Calculate the frequency of the signal.

Three complete waves in 2·5 seconds:

$$f = \frac{N}{t} = \frac{3}{2 \cdot 5} = 1 \cdot 2 \text{ Hz}$$

TASK Quick questions

1. The following diagram represents a wave travelling from left to right.

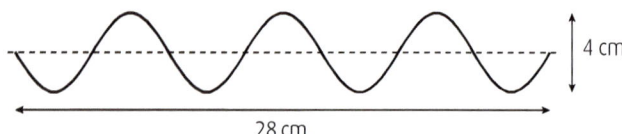

   Determine the wavelength and amplitude of the wave.

2. A person standing on a jetty by the sea counts eight waves passing a marker in 24 seconds. Calculate the frequency of the waves.

3. An oscilloscope displays an electronic signal.

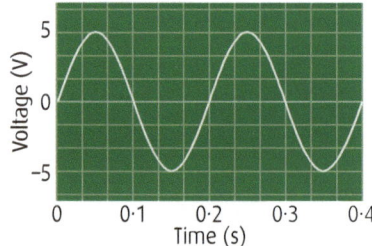

   (a) What is the frequency of the signal?
   (b) What is the amplitude of the signal?

SPEED, FREQUENCY AND WAVELENGTH

Speed v, wavelength λ and frequency f are related to each other in calculations whenever waves are present – for example, in sound waves, water waves and light waves.

The **wave equation** connects v, f and λ:

$$v = f\lambda$$

Rearranged versions of the wave equation are:

$$f = \frac{v}{\lambda}$$

$$\lambda = \frac{v}{f}$$

where v = **wave speed** in metres per second (m s^{-1}), f = **frequency** of wave in Hertz (Hz) and λ = **wavelength** of wave in metres (m).

A buzzer produces sound waves with a frequency of 850 Hz and a wavelength of 0·4 m.

Calculate the speed of the waves.

$$v = f\lambda$$

$$v = 850 \times 0 \cdot 4 = 340 \text{ m s}^{-1}$$

EXAMPLE

A seagull is floating on the sea close to the shore.

Calculate the frequency with which the seagull bobs up and down when waves of wavelength 12 m and speed 2·4 m s^{-1} pass by.

ANSWER

$$f = \frac{v}{\lambda} = \frac{2·4}{12} = 0·2 \text{ Hz}$$

WAVE SPEED

Wave speed can also be calculated using the relationship:

$$\textbf{speed} = \frac{\textbf{distance}}{\textbf{time}} \left(v = \frac{d}{t} \right)$$

Rearranged versions:

$$d = vt \qquad\qquad t = \frac{d}{v}$$

EXAMPLE

Sound waves travel through water at a speed of 1500 m s^{-1}.
A whale makes sounds underwater to communicate with other whales.

Calculate the distance travelled by the sounds in water in a time of 75 s.

ANSWER

$$d = vt = 1500 \times 75 = 112\ 500 \text{ m}$$

EXAMPLE

A bicycle bell produces a single note in air with a frequency of 850 Hz.

The wave speed is 340 m s^{-1}.
Calculate the wavelength of the sound waves produced.

ANSWER

$$\lambda = \frac{v}{f} = \frac{340}{850} = 0·4 \text{ m}$$

EXAMPLE

A student drops a stone into a pond and measures a time of 12 s for the crest of a wave to travel 8 m.

Calculate the speed of the wave.

ANSWER

$$v = \frac{d}{t} = \frac{8}{12} = 0·67 \text{ m s}^{-1}$$

EXAMPLE

Calculate the time taken for the sound of a thunderstorm to travel a distance of 15 300 m in air if the speed of sound in air is 340 m s^{-1}.

ANSWER

$$t = \frac{d}{v} = \frac{15\ 300}{340} = 45 \text{ s}$$

Wave speed can be calculated using $v = f\lambda$ or $v = \frac{d}{t}$.

Sometimes both relationships have to be used in one question.

DON'T FORGET

To calculate v, sometimes the equation $v = f\lambda$ is used and sometimes the equation $v = \frac{d}{t}$.

THINGS TO DO AND THINK ABOUT

1. Waves of wavelength 0·05 m travel across a pond at a frequency of 6 Hz.
   Calculate the wave speed.
2. During earthquakes, P-type waves are produced that travel through the Earth.
   Calculate the frequency of such waves during one earthquake if their wavelength is 10 m and their speed is 6000 m s^{-1}.
3. Sound waves of frequency 800 Hz travel through a length of aluminium. The speed of sound in aluminium is 5200 m s^{-1}.
   Calculate the wavelength of the sound waves in aluminium.
4. Water waves move a distance of 12 m in 4·8 s.
   Calculate the wave speed.
5. Calculate the time taken for water waves travelling at 8 m s^{-1} to travel 800 m.
6. Calculate the distance travelled by sound waves in air in 8 s if the speed of sound in air is 340 m s^{-1}.

SOUND 1

USING AN OSCILLOSCOPE

When a microphone is connected to an oscilloscope (an instrument that converts electrical signals into wave patterns on a screen), the pattern of sound waves can be examined.

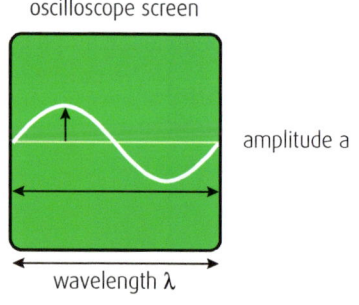

oscilloscope screen

amplitude a

wavelength λ

An oscilloscope screen.

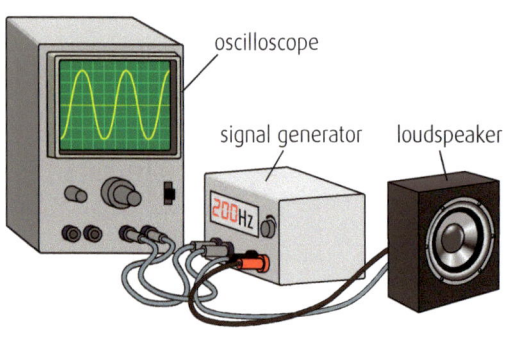

oscilloscope

signal generator

loudspeaker

200Hz

A signal generator and loudspeaker connected to an oscilloscope.

A signal generator produces electrical signals. The frequency and amplitude of these signals can be increased or decreased. An oscilloscope and loudspeaker can be connected to a signal generator to investigate how sound frequency and loudness change.

When the frequency and loudness of a sound are adjusted, the patterns on the oscilloscope change, as shown in the diagrams on the right.

original pattern

increased **loudness**   increased **frequency**   increased **frequency** and **loudness**

The wave has larger amplitude

More waves

More waves and larger amplitude

Effects of changing the frequency and loudness of sounds.

ANALYSING SOUND WAVES

Human speech can be converted into electrical signals. These signals can be analysed and used in a variety of useful processes.

Voice recognition software

Computers can be used to analyse the electrical signals converted from a person's speech and to convert these signals into text. This process is useful for people who may have some discomfort or difficulty writing or typing, or simply to manage bank accounts by telephone.

MEASURING THE SPEED OF SOUND IN AIR

The speed of sound in air can be measured by timing how long sound takes to travel over a certain distance, and then using $v = \dfrac{d}{t}$.

Measuring the speed of sound in the laboratory

An electronic timer that can accurately measure small time intervals is required. Over short distances (such as in a laboratory experiment), sound travels too quickly in air to be measured by a hand-operated stop watch.

Two microphones, separated by a measured distance, are used to start and stop an electronic timer.

A sharp sound is made on one side of the microphones.

As the sound passes microphone A, the timer starts, and then stops timing when the sound reaches microphone B.

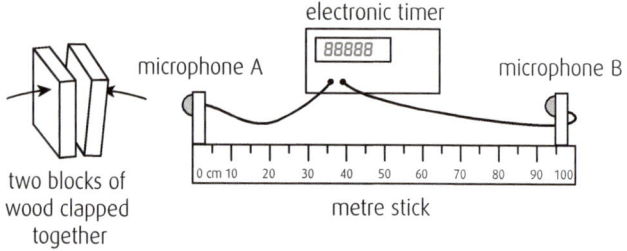

electronic timer

88888

microphone A

microphone B

two blocks of wood clapped together

0 cm 10   20   30   40   50   60   70   80   90 100

metre stick

Experimental set-up to measure the speed of sound in air.

The values for the distance between the microphones and the reading displayed on the timer are then used in the equation $v = \dfrac{d}{t}$ to calculate the speed of sound.

Sample readings: distance between microphones = 0·85 m; timer reading = 0·0026 s.

$$v = \frac{d}{t} \quad v = \frac{0·85}{0·0026} = 327 \text{ m s}^{-1}$$

This is the experimental value calculated for the speed of sound after completing the experiment once.

It is good experimental practice to repeat the experiment and to alter the distance between the microphones each time to obtain several results for the speed of sound, which can then be averaged. This leads to a more reliable result.

The accepted value for the speed of sound in air for calculations in N4 (and N5) Physics is 340 m s⁻¹.

This experimental method can also be used to measure the speed of sound in liquids or solids (e.g. metals).

MEASURING THE SPEED OF SOUND IN LIQUIDS

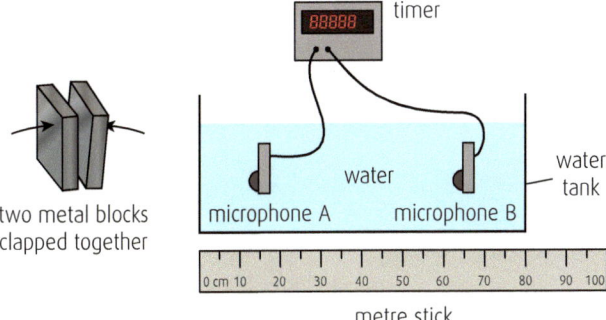

Experimental set-up to measure the speed of sound in liquids.

Sample readings: distance between microphones = 0·58 m; timer reading = 0·0004 s.

$$v = \frac{d}{t} \quad v = \frac{0·58}{0·0004} = 1450 \text{ m s}^{-1}$$

The accepted value for the speed of sound in water for calculations in N4 (and N5) Physics is 1500 m s⁻¹.

MEASURING THE SPEED OF SOUND IN SOLIDS

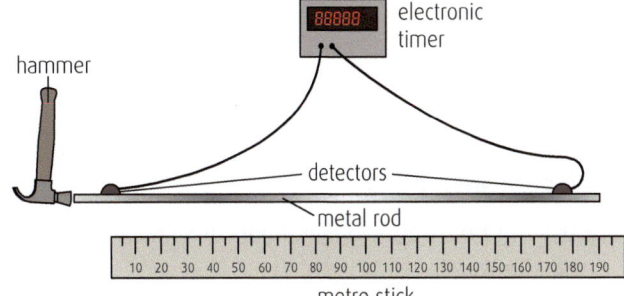

Experimental set-up to measure the speed of sound in solids.

A timer is connected to a hammer and a detector. The detector is attached to one end of a metal rod.

A hammer is used to strike one end of the metal rod. The timer starts timing when the hammer makes contact.

The timer stops timing when the sound travels through the rod and reaches the detector.

Sample readings: length of steel rod = 0·75 m; timer reading = 0·00015 s.

$$v = \frac{d}{t} \quad v = \frac{0·75}{0·00015} = 5000 \text{ m s}^{-1}$$

The accepted value for the speed of sound in steel for calculations in N4 (and N5) Physics is 5200 m s⁻¹.

 **DON'T FORGET**

To calculate the speed of sound in a solid, liquid or gas, the distance travelled and the time taken for the sound to travel this distance are used with the equation $v = \dfrac{d}{t}$.

 THINGS TO DO AND THINK ABOUT

The following patterns were obtained on an oscilloscope when four sounds were detected by a microphone connected to it.

The oscilloscope controls were not altered between sounds.

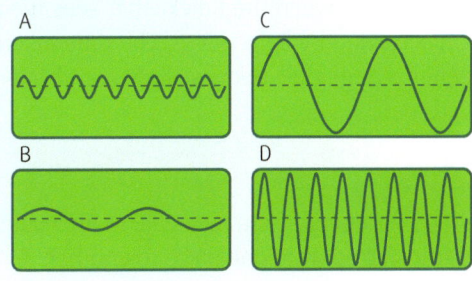

Which pattern represents:

(a) a low frequency, loud sound?
(b) a high frequency, quiet sound?

SOUND 2

MEASURING THE SPEED OF SOUND IN AIR USING LONGER DISTANCES

An experiment to measure the speed of sound can be carried out in a large field. The time taken for a sound to travel a longer distance can be measured using a hand-held stopwatch.

Two students stand at **opposite** ends of the field.

The distance between the students is measured using a trundle wheel or long tape measure.

One student holds up two metal bars, which can be seen by the other student who has the stopwatch.

The student with the stopwatch starts timing when she sees the metal bars being banged together, and then stops timing when she hears the sound of the metal bars.

The time on the stopwatch is the time taken for the sound to travel the distance between the students.

The experiment should be repeated for different distances and an average value for the speed of sound obtained.

Some sample readings for this experiment are:

Set 1:
Distance = 115·5 m; time = 0·35 s
$$v = \frac{d}{t} \quad v = \frac{115\cdot5}{0\cdot35} = 330 \text{ m s}^{-1}$$

Set 2:
Distance = 135·2 m; time = 0·40 s
$$v = \frac{d}{t} \quad v = \frac{135\cdot2}{0\cdot40} = 338 \text{ m s}^{-1}$$

Set 3:
Distance = 140·7 m; time = 0·42 s
$$v = \frac{d}{t} \quad v = \frac{140\cdot7}{0\cdot42} = 335 \text{ m s}^{-1}$$

160m

Measuring the speed of sound over longer distances.

The average of these results can be calculated:

$$v_{average} = \frac{330 + 338 + 335}{3} = 334 \text{ m s}^{-1}$$

In this experiment, the stopwatch is started when the timekeeper sees the rods being banged together. This information is carried by light waves. The time taken for light to travel the distance to the timekeeper is negligible because light travels so much faster than sound.

SOUND IN A VACUUM

Sound waves rely on particles to transfer energy from molecule to molecule. Without these particles, energy cannot be transferred from one place to another.

If there are no particles inside a container to transfer this energy, then sound will not travel through the container.

Experiment to demonstrate that sound cannot travel through a vacuum

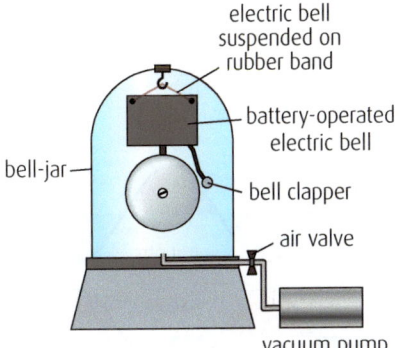

electric bell suspended on rubber band

battery-operated electric bell

bell-jar

bell clapper

air valve

vacuum pump

Measuring the speed of sound in a vacuum.

A battery-operated bell is suspended inside a bell-jar. The bell-jar is connected to a vacuum pump.

The bell is switched on and the ringing sound can be heard.

The vacuum pump is then switched on to remove air from inside the jar.

When all of the air has been removed, the valve is closed and the vacuum pump is switched off.

The ringing sound can no longer be heard, but the bell clapper can be seen to be moving.

The sound waves cannot travel because there are no air particles inside the bell-jar to transfer the sound energy.

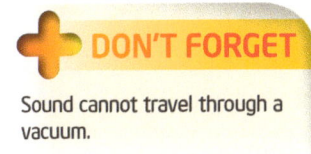

DON'T FORGET

Sound cannot travel through a vacuum.

SOUND LEVELS

Noise pollution (or environmental noise) can be described as unpleasant levels of sound. Noise pollution can be created by humans, animals or machines. Many of these sounds are unpleasant because they are too loud for comfort; some may even damage our hearing.

Common sources include noise from road and air traffic, concerts, building sites and noisy neighbours.

Potential sources of noise pollution.

Sound level meters are used to measure sound levels. It is important to measure and monitor noise levels as prolonged exposure can damage our hearing and even cause deafness.

Sound levels are measured in **decibels** (dB). Hearing damage can be caused by exposure to sound levels above 90 dB.

People who are regularly exposed to sound levels greater than 90 dB should wear ear protectors to absorb the sound energy. There are many regulations to protect people from excessive noise. The **threshold of hearing** is 0 dB. This is the quietest sound that a young person with undamaged hearing can detect.

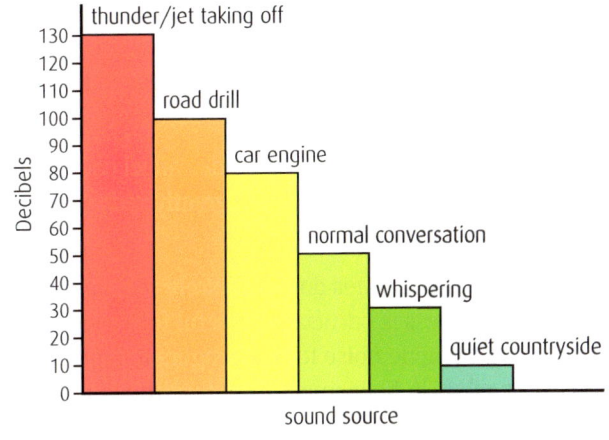

Decibels

thunder/jet taking off

road drill

car engine

normal conversation

whispering

quiet countryside

sound source

A range of sound levels.

THINGS TO DO AND THINK ABOUT

1. In an experiment, the time taken for sound waves to travel 2·4 m was 0·0075 s.
   Calculate the speed of sound from these results.

2. In a repeated experiment to measure the speed of sound in air, the following results were obtained: 330, 328, 334, 336, 338 and 333 m s^{-1}.
   Calculate the average speed obtained in the experiment.

3. During an approaching storm, a person hears thunder 2·4 s after a lightning flash. The speed of sound in air is 340 m s^{-1}.
   Calculate the person's distance from the storm.

SOUND 3

REDUCING THE EFFECT OF LOUD NOISES

Sound is transferred as waves of energy. There are many examples of excessive sound levels being reduced for our comfort and safety. Methods include the use of materials that absorb sound energy and reduce the amount of sound passing through. Unwanted noise can be reduced by using these energy-absorbing materials.

1. Building regulations provide guidance on the amount of **sound insulation** that should be included in houses to protect the occupants from outside noise.

2. **Ear protectors** contain material that absorbs sound energy and thus reduces the amount of energy reaching the wearer's ears.

3. **Noise-cancelling headphones** are designed to reduce the amount of repetitive noise that reaches the wearer's ears.

Noise-cancelling headphones

These headphones have normal sound-insulating material to reduce the amount of sound energy that penetrates to the ear. In addition, the headphones contain microphones to detect incoming sound.

An electronic circuit inside the headphones re-creates sound with the same loudness and frequencies as the outside sounds, but which produces wave crests and troughs inside the headphones that are out of sequence with the outside sound waves. Wave troughs are produced inside the headphones when crests are arriving outside.

The result is that the two sets of sounds cancel each other out and the wearer hears only silence (or at least greatly reduced sound). This property of waves is known as **interference**.

In practice, the headphones do not entirely reduce the outside noise to zero. However, the process is particularly useful for reducing repetitive sounds, such as aircraft engine noise for passengers in flight. These headphones are also used by pilots in helicopters to reduce the loud sounds from the rotors and engine.

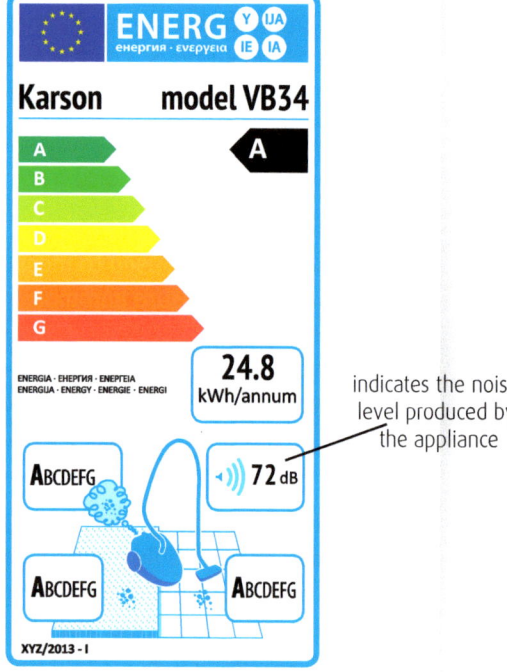

sound from outside headphones

sound created inside headphones

sound waves have "cancelled" each other

How noise-cancelling headphones work.

Appliance label showing noise level produced.

NOISE LEVELS PRODUCED BY DOMESTIC APPLIANCES

European Union law now insists that manufacturers must label domestic appliances with the noise level emitted by the product where noise is a relevant issue for the appliance. This is to assist consumers in deciding which appliance to buy.

 **TASK**

Investigate the labels produced by manufacturers for various household appliances. Carry out research into which household appliances produce the most noise when in use. Present your results in a table.

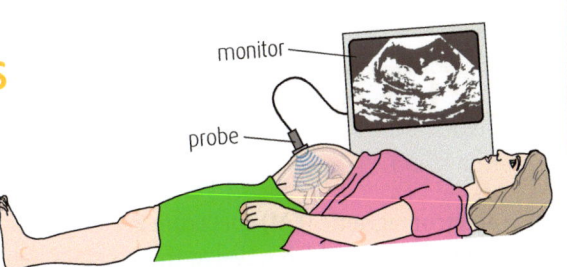

PRACTICAL APPLICATIONS OF SOUND WAVES

Ultrasound and ultrasonic scanning

High frequency vibrations beyond the frequency of human hearing (above 20 000 Hz) are called **ultrasound**.

An example of the use of ultrasound is in medicine. Images of internal organs and tissues are created by sending ultrasonic waves into the body using a probe. When they pass from one type of tissue to another, some waves are reflected back to a receiver in the probe. A computer is used to analyse the return times of the reflected waves to produce an image of the inside of the body. Ultrasound scans are regularly used to monitor the development of unborn babies and are much safer than using X-rays, which may be harmful.

Ultrasound scans are used to monitor the development of unborn babies.

EXAMPLE

Ultrasound is used by doctors for both treatment and diagnosis. Pulses of ultrasound are used to produce local heating of muscles deep inside the body. The heating effect can help relieve pain in the muscles.

(a) What is meant by ultrasound?
(b) Calculate the time required for a pulse of ultrasound to travel through 3 cm of muscle (the speed of sound in muscle is 1500 m s⁻¹).

ANSWER

(a) Sound waves with frequencies greater than 20 000 Hz or sounds above the frequency range that can be heard by humans.

(b) $v = \dfrac{d}{t} = 1500 = \dfrac{0\cdot03}{t}$

$t = \dfrac{0\cdot03}{1500} = 0\cdot00002$ s.

➕ DON'T FORGET

Ultrasound waves are high frequency sound waves that travel at the same speed as the sounds humans can hear.

Sound waves are also used in this way to determine the depth of the sea bed when constructing charts for navigation.

EXAMPLE

A ship is mapping the sea bed using ultrasound waves. When stationary, the ship transmits pulses of ultrasound waves down to the sea bed and receives back the reflected pulses.

The transmitted ultrasound waves have a frequency of 40 000 Hz.

One pulse of ultrasound is received back at the ship 0·38 s after being transmitted.

Calculate the depth of the sea bed (the speed of sound in water is 1500 m s⁻¹).

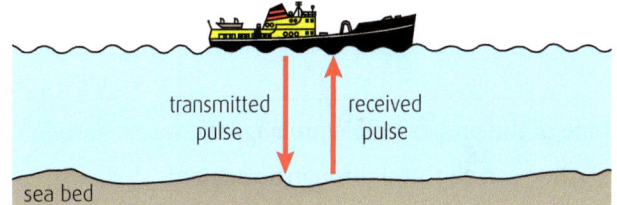

transmitted pulse   received pulse

sea bed

Using ultrasound to measure the depth of the sea bed.

ANSWER

Time taken for sound to travel one way to sea bed only $= \dfrac{0\cdot38}{2} = 0\cdot19$ s

$d = v \times t = 1500 \times 0\cdot19 = 285$ m

THINGS TO DO AND THINK ABOUT

1. What units are used for measuring sound levels?
2. What is the danger level for sounds to be harmful to human hearing?
3. Give an example of a sound that exceeds the danger level for humans.
4. How are sound levels designed to be reduced for occupants during the construction of houses?
5. Sound produced by a whale underwater in the Antarctic was detected 8900 km away in the Atlantic Ocean.
   Calculate the time taken for the sound to travel this distance (speed of sound in water = 1500 m s⁻¹).

ELECTROMAGNETIC SPECTRUM 1

The **electromagnetic (EM) spectrum** is a large family of waves with a wide range of wavelengths, all of which travel at the speed of light (300 000 000 or 3×10^8 m s^{-1} in a vacuum). This is also the speed used for the electromagnetic spectrum wave speed when travelling through air.

However, each member of the family has different wavelengths, frequencies and energies.

QUICK FACTS ABOUT THE ELECTROMAGNETIC SPECTRUM

High frequency Short wavelength Greatest energy						Low frequency Long wavelength Least energy
EM spectrum	**Gamma rays**	**X-rays**	**Ultraviolet**	**Visible light**	**Infra-red (heat rays)**	**Radio waves**
Example of source	Radioactive materials	X-ray scanner	Ultraviolet lamp	Visible spectrum of colour	Heater	Radio transmitter
Example of detector	Geiger–Müller tube	Causes fogging in a photographic plate	Causes some symbols in genuine banknotes to glow	Eyes	Thermal imaging camera	Microwaves — Mobile phone aerial; TV waves — TV aerial; Radio waves — Radio aerial

Some of the properties electromagnetic waves have in common are:

- they can all travel through a vacuum;
- they all travel at the same speed through a vacuum (or air) (300 000 000 m s^{-1});
- they are all waves;
- they all transfer energy.

Applications and hazards

Each part of the electromagnetic spectrum has properties that are useful in different applications. However, there are also hazards associated with each part of the spectrum that have to be carefully managed.

COMMON SOURCES, APPLICATIONS AND HAZARDS OF ELECTROMAGNETIC RADIATION

The different types of radiation in the electromagnetic spectrum have many applications. The Sun and stars are sources of all the wavelengths of radiation present in the electromagnetic spectrum. However, the most common applications involve sources that are produced on Earth.

DON'T FORGET

The electromagnetic spectrum consists of a range of waves that transfer energy from one place to another.

	Typical source	Detector	Applications and uses	Additional facts	Hazards
Gamma rays	Radioactive substances, nuclear reactors	Geiger–Müller tube and counter	Radioisotopes that emit gamma rays are used as tracers in medicine to investigate blood flow and related problems in humans using a gamma camera detector Sterilisation of medical equipment Scanning shipping containers at ports	Highest energy of electromagnetic spectrum Cannot penetrate Earth's atmosphere from outer space	Accidental exposure can kill human cells Can cause cancer

	Typical source	Detector	Applications and uses	Additional facts	Hazards
X-rays	X-rays are produced when very fast-moving electrons collide with a metal target X-ray machines	X-rays darken photographic films X-ray image intensifiers Geiger–Müller tube and counter	Medical imaging, giving three-dimensional images of the internal body structure, used especially in the diagnosis of broken bones Analysis of atomic structures in X-ray crystallography Airport security scanners	High-energy waves Cannot penetrate Earth's atmosphere from outer space	Accidental (or too much) exposure can kill human cells Can cause cancer
Ultraviolet radiation	The Sun is an important source of ultraviolet radiation Ultraviolet lamps, including mercury vapour lamps	Exposure to ultraviolet radiation causes some materials to fluoresce, i.e. to glow An ultraviolet photodiode is an electronic device which detects ultraviolet radiation	Ultraviolet radiation causes a chemical reaction in the skin that produces the important nutrient vitamin D Ultraviolet radiation is used in the treatment of certain skin conditions Disinfection of hospital equipment Used by dentists to 'cure' or harden composite material used for fillings	Most ultraviolet radiation from the Sun is absorbed in the upper atmosphere by the ozone layer The ultraviolet radiation range is sometimes separated into three bands: UVA, UVB and UVC UVC has the highest frequency (hence the highest energy) and is filtered out by the ozone layer in the Earth's atmosphere	Accidental exposure can cause eye damage Overexposure can cause skin cancer UVA and UVB radiation cause sunburn and tanning; overexposure can cause skin cancer
Visible light	The Sun is the primary source of visible light Light bulbs, lasers	Photodiodes Phototransistors Light-dependent resistors	Light from lasers is widely used in communication through optical fibres Supermarket checkout readers use laser light to scan barcodes for price information	The eye responds to visible light, which occupies the smallest range of wavelengths in the electromagnetic spectrum Red light has the longest and blue light the shortest wavelength in the visible region	Intense light sources can damage the retina if viewed too closely Laser light is particularly intense and can cause damage to eyesight
Infra-red rays (heat rays)	Infra-red radiation is received from the Sun Infra-red heaters	Black bulb thermometer Thermopile	Used as a heating source Humans emit infra-red radiation – night thermal imaging infra-red cameras are used to detect infra-red radiation in the dark and to locate casualties trapped in collapsed buildings Used in medical diagnosis and treatment Passive infra-red detectors are used in intruder alarms	Infra-red rays (heat rays) are responsible for heat transfer by radiation Over half of the radiation received on Earth from the Sun is infra-red radiation Most of the emitted energy from the Earth through the atmosphere back into space is in the form of infra-red radiation	Intense heat can cause burns to the skin Too much exposure can cause sunburn
Microwaves	Microwave ovens produce waves with a wavelength of approximately 12 cm	Radar detector dishes	Used extensively for communication – for example, in satellite phones and television outside broadcasts Used in medicine to destroy (ablate) cancerous tumours	Limited range in built-up areas	Exposure to high-energy microwaves can cause internal heating of body tissues
Radio waves	Radio waves are produced by various transmitters	TV aerials Radar detector dishes	Used extensively for communication Transmitters connected to amplifiers produce radio waves		Exposure to high-energy radio waves can cause internal heating of body tissues

THINGS TO DO AND THINK ABOUT

1. What is the speed of the waves of the electromagnetic spectrum family in air?
2. Which types of radiation are B and D in the following table?

Gamma rays	B	Ultraviolet radiation	D	Infra-red radiation (heat rays)	Radio waves

3. What detects microwaves inside a mobile phone?
4. Which waves have the highest frequency?
5. Which waves have the lowest energy?

ELECTROMAGNETIC SPECTRUM 2

RADIO WAVES

Radio waves have the lowest energy of the different types of radiation in the electromagnetic spectrum.

However, exposure to high levels of radio waves is dangerous. Microwave ovens produce microwaves with great energy. Householders are advised to ensure that microwave ovens do not leak this radiation by keeping door seals clean and checking for signs of damage to the oven.

INFRA-RED RADIATION

Heat rays from the Sun reach the Earth and raise its temperature. Surfaces absorb this heat energy when facing the Sun and radiate it back to the atmosphere when in darkness. The Earth's average temperature is maintained by this process. However, the production of carbon dioxide (CO_2) gas by human activities (mainly from burning fossil fuels) has caused an increase in the amount of CO_2 present in the atmosphere. This prevents infra-red radiation from escaping and is causing the planet's average temperature to gradually increase. Governments are trying to reduce CO_2 emissions to stop this 'global warming' effect, which may lead to changes in the Earth's climate.

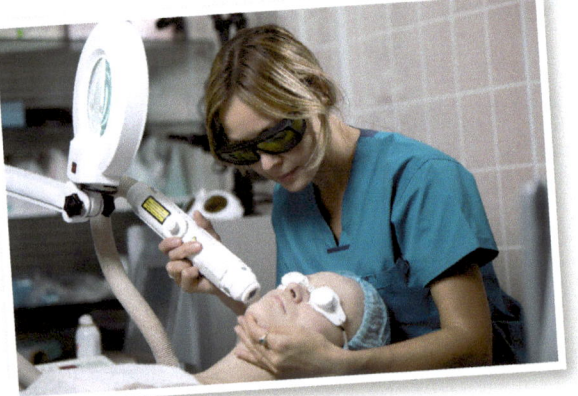
The door seals on microwave ovens should be checked regularly to prevent leakage of microwave radiation.

VISIBLE LIGHT

Our eyes naturally blink when we are exposed to very bright light, which can cause damage to the sensitive retina at the back of the eye.

Laser light, however, is very intense and can damage the retina without being noticed. Lasers used as bar-code readers in supermarkets are low power and should not cause eye damage through accidental viewing. The power of laser light sources is strictly managed to ensure that powerful lasers are only used by trained operators wearing eye protection. People are advised never to look directly at laser light.

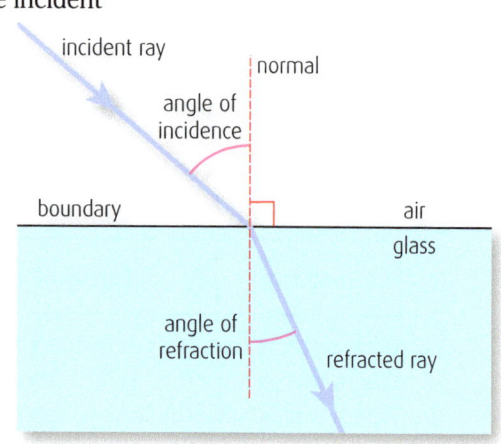
Lasers can be used safely for cosmetic skin treatments.

Refraction of light is the change of speed of light when it travels from one material into another.

For example when a ray of light travels through air into a block of glass, the incident (approaching) ray makes an angle of incidence with the normal.

The normal is a dotted line drawn at right angles to the boundary.

Similarly, the refracted ray (inside the glass) makes an angle of refraction with the normal.

The angle in glass is always less than the angle in air.

When the incident ray travels in the same direction of the normal, there is no change in direction in the glass.

This property of refraction is used in spectacle and contact lenses to change the direction of light when it enters the eye, to improve long and short sight.

incident ray

normal

angle of incidence

boundary   air

glass

angle of refraction

refracted ray

PROTECTION FROM ELECTROMAGNETIC RADIATION

Many of the applications of the electromagnetic spectrum help to improve the living standards, health and wellbeing of humans.

However, it is important to protect ourselves against the dangers of exposure to harmful radiation.

The Earth's atmosphere protects us by absorbing gamma radiation, X-rays and most of the ultraviolet radiation received from space.

Whenever harmful electromagnetic radiation is produced on Earth, precautions must be taken to prevent accidental exposure to humans and other living organisms.

ULTRAVIOLET RADIATION

Some ultraviolet radiation from the Sun does penetrate the atmosphere. Overexposure to ultraviolet radiation can cause skin cancer. Sunbathers are always warned not to stay in sunlight for too long and to always wear protective sun cream.

The **ozone layer** in the Earth's atmosphere acts as a filter for harmful ultraviolet radiation. Scientists monitor this layer. It has been damaged by gases used in refrigerants and as propellants in aerosols and increased levels of ultraviolet radiation have been detected on the Earth's surface. Governments have reduced the amount of emissions allowed and some have even banned the use of these gases to protect the ozone layer from developing further holes.

DON'T FORGET

The different types of radiation in the electromagnetic spectrum are very useful to humans, but safety precautions must always be taken when using them.

X-RAYS

A wall or a thick glass window is required to protect the operators of X-ray machines; operators always move away from the machines when they are switched on.

GAMMA RADIATION

Some radioactive materials – for example, the radioactive waste from nuclear power stations – emit gamma radiation. This waste remains dangerous for many years and has to be stored carefully. Several centimetres of lead and thick concrete are required to absorb the energy from gamma radiation.

Radioactive waste from nuclear power stations remains active for many years.

THINGS TO DO AND THINK ABOUT

1. Investigate how concave and convex lenses change the direction of light rays. Make drawings of how each lens does this. Which lenses are used to correct (a) short sightedness, (b) long sightedness?
2. Name a detector of gamma radiation.
3. Describe one hazard associated with X-rays.
4. Describe one useful application of gamma radiation.
5. Name one detector of infra-red radiation.
6. Which types of radiation from space are mostly filtered by the Earth's atmosphere?
7. Describe the problem with radioactive waste from nuclear power stations.
8. Describe one use of ultraviolet radiation.

NUCLEAR RADIATION 1

STRUCTURE OF ATOMS

Nuclear radiation is energy that comes from the nucleus of an **atom**. An atom consists of a central nucleus containing **protons** and **neutrons**. **Electrons** orbit the nucleus.

Name	Charge (units)	Mass (units)
Proton	Positive +1	1
Neutron	Neutral 0	1
Electron	Negative –1	1/2000 (negligible)

SOURCES OF RADIOACTIVE MATERIALS

There are both natural and artificial sources of radioactive materials. The artificial sources are made by humans.

Naturally occurring radioactive materials

Radon is a radioactive gas found in rocks (such as granite) and soils.

Radioactive versions of some elements (e.g. potassium-40) are present in very small amounts in food. Coal contains small amounts of naturally occurring radioactive uranium.

Artificial radioactive materials

A version of the element technetium, called technetium-99m, is produced in nuclear reactors for use in medical procedures.

Naturally radioactive uranium can be modified to produce an enriched version (known as uranium-235). Fuel pellets containing uranium-235 are used in fuel rods in nuclear power stations.

Model of an atom.

TYPES OF RADIATION

There are three types of nuclear radiation: alpha, beta and gamma radiation.

Nuclear radiation	Symbol	Form	
Alpha	α	**Alpha (α) radiation** — alpha particle — 2 neutrons, 2 protons	A large positive (helium) nucleus
Beta	β	**Beta (β) radiation** — beta particle = 1 electron	A fast-moving electron
Gamma	γ	**Gamma (γ) radiation** — gamma ray = energy wave	An electromagnetic wave

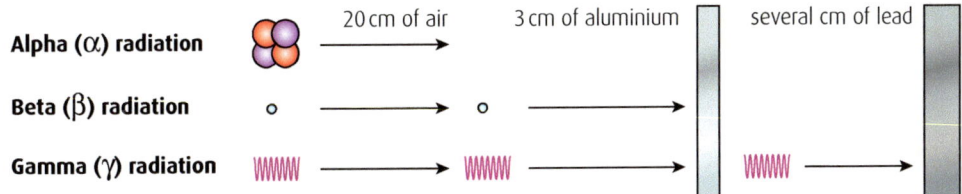

Alpha (α) radiation   20 cm of air   3 cm of aluminium   several cm of lead

Beta (β) radiation

Gamma (γ) radiation

Different types of radiation penetrate materials by different amounts.

ABSORPTION OF RADIATION

It is vital to know how far radiation can travel (or penetrate) before it is absorbed and is no longer dangerous. This knowledge allows us to select materials to shield humans from radiation.

Alpha radiation can only travel a few centimetres in air before it is absorbed. Beta and gamma radiation can travel further through air. A thin sheet of paper or 20 cm of air can absorb alpha radiation. Beta radiation can be absorbed by 3 or 4 cm of aluminium, but gamma rays require several centimetres of lead to absorb most of their energy.

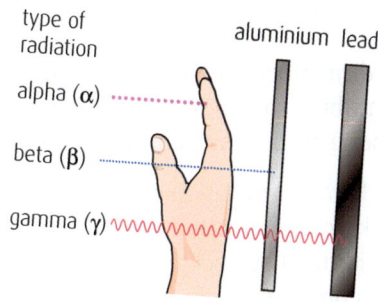

type of radiation   aluminium   lead

alpha (α)

beta (β)

gamma (γ)

Different types and thicknesses of material are required to absorb α, β and γ radiation.

TASK   Quick questions

1. State the names of the two types of particle found in the nucleus of an atom.
2. State the charge of each of these particles.
3. State the name and charge of the particle that orbits the nucleus of an atom.
4. Name one natural source of radiation.
5. Name one artificial source of radiation.
6. State the names and write the symbols of the three types of nuclear radiation.
7. Which type of radiation can be absorbed by a thin sheet of paper?
8. Which type of radiation can penetrate most materials?

A film badge measures exposure to radiation.

RADIATION DETECTORS

Geiger–Müller tubes

A Geiger–Müller tube can be used to detect radiation. It is usually connected to a counter that counts the number of atoms that have decayed.

Each time an atom decays, the Geiger–Müller tube detects it and records it on the counter.

Film badges

We need to monitor the amount of radiation absorbed by people who work with radiation over periods of time to ensure that safety limits of exposure are not exceeded. These workers wear film badges.

Radiation can cause fogging of photographic film. When the film inside the badge is developed, the amount of fogging gives an indication of what type of, and how much, radiation the badge has been exposed to.

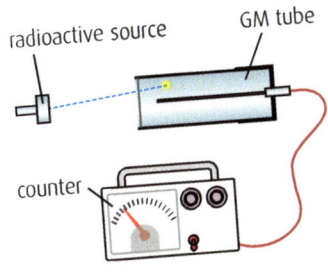

radioactive source   GM tube

counter

A Geiger–Müller tube and counter.

DON'T FORGET

When radioactive substances decay, they emit alpha, beta or gamma radiation.

THINGS TO DO AND THINK ABOUT

Carry out research into other types of radiation detectors, such as cloud chambers and spark counters.

NUCLEAR RADIATION 2

BACKGROUND RADIATION

Radioactive materials exist everywhere, so there is always radiation around us. This is called background radiation. There are various sources of radiation that combine to cause background radiation.

Medical procedures such as X-rays and radiotherapy contribute to the exposure of humans to radiation. There are other artificial and natural sources of background radiation.

Background radiation can be measured locally. The level of background radiation depends on where you live. Humans are not harmed by background radiation because it has always been present all around us.

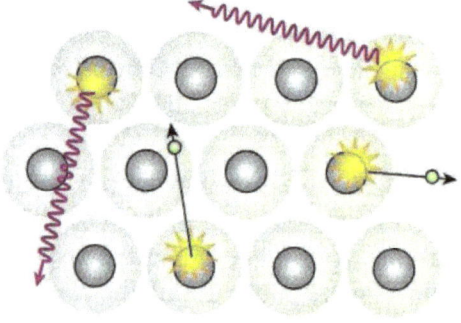

Sources of background radiation.

RADIATION EMITTED FROM THE ATOMS OF RADIOACTIVE ELEMENTS

When an atom of a radioactive element emits nuclear radiation, the atom may change into a new form of the same element, or change to become a different element.

The atom is said to have **decayed** when it has emitted radiation.

Activity

When a radioactive source emits radiation, the radioactive atoms of the source decay.

The **activity** of a radioactive substance is the number of atoms that decay each second and is measured in **becquerel (Bq)**.

One becquerel (Bq) = one decayed atom per second.

Radioactive decay of an atom.

HALF-LIFE

Radioactive materials are used in homes, hospitals, nuclear power stations and in industry.

It is important to know how long these substances will remain radioactive. This will allow precautions to be taken to protect radiation workers and the general public.

The **half-life** of a radioactive substance is a useful measurement and is important for the safe management of radioactive materials.

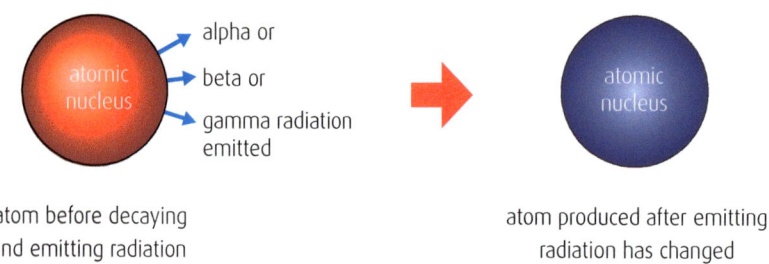

atom before decaying and emitting radiation

atom produced after emitting radiation has changed

Decay of an atom of a radioactive element.

The half-life of a radioactive element is the time taken for the **activity** of the element to decrease to half of its original value.

The decay of radioactive atoms is a **random** process. This means that, out of the millions of atoms in any sample of radioactive material, it is not possible to predict exactly when one particular atom will decay and emit radiation.

However, because of the huge number of atoms in any sample of a radioactive element, at any instant, there will be many atoms decaying. The number of decaying atoms can be detected and the time measured.

This allows the half-life of the substance to be determined.

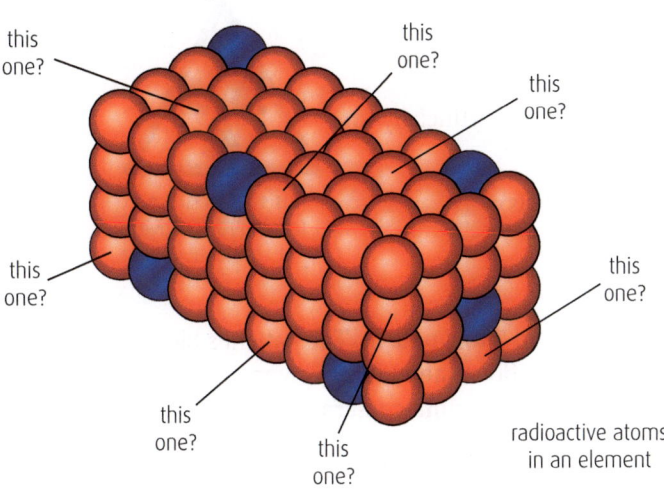

Radioactive decay: which atom will decay next?

EXAMPLE

Uranium-235 has a half-life of 700 million years.
Uranium-235 is used in reactors in nuclear power stations. The fuel used in reactors produces the heat energy used to generate electrical energy.

Carbon-14 has a half-life of 5730 years.
Carbon-14 is found in organic materials. It is used in radiocarbon dating to find out the age of geological and archaeological materials.

Technetium-99 has a half-life of six hours
Technetium-99 is commonly used in medicine as a tracer to monitor blood flow.

THINGS TO DO AND THINK ABOUT

1. What does an atom emit when it decays?

2. What is the activity of a radioactive substance?

3. What is the half-life of a radioactive substance?

APPLICATIONS AND USES OF NUCLEAR RADIATION

MEDICAL USES OF NUCLEAR RADIATION

Radioactive materials are sometimes produced artificially for use in medical procedures to:

- find out about health problems (**diagnosis**);
- to improve health (**treatment**).

This symbol is used to label radioactive material.

Diagnosis

A radioactive liquid called a **tracer** is injected into a patient. The patient's blood carries this liquid around the body. The radiation (gamma rays) given off by the radioactive liquid in the blood is detected by a **gamma camera**.

Areas where the radiation detected is high or low indicate places where there might be a problem.

The gamma camera detects gamma rays using a crystal. This produces flashes of light that are converted into electrical pulses and are analysed by a computer. An image is displayed on a monitor.

Radiation used in diagnosis.

Treatment

Radiotherapy is the use of radiation to treat some medical conditions (such as cancer).

Radiotherapy uses a machine that emits gamma radiation. The gamma radiation kills cancer cells, but only damages healthy cells, which are then able to recover.

The machine emitting the gamma rays is rotated around the patient. This directs the beam to the tumour from different angles so that the healthy tissues surrounding the tumour do not receive too much radiation.

Radiation treatment is used in medicine.

INDUSTRIAL USES OF NUCLEAR RADIATION

Nuclear radiation is often used in industry to allow complex processes to be completed quickly and easily.

Sterilisation of food

Some foods are subjected to gamma radiation to kill bacteria or insects that may be present. This allows the food to last for longer before becoming unfit to eat. The gamma radiation can even be used after the food has been packaged because gamma rays can penetrate most materials. Some people are against the use of this process because they say that it changes the taste of the food.

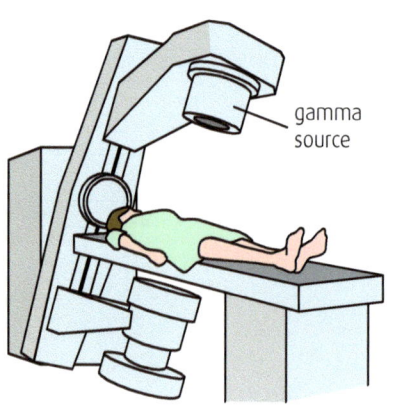

Radiotherapy is used to treat cancer.

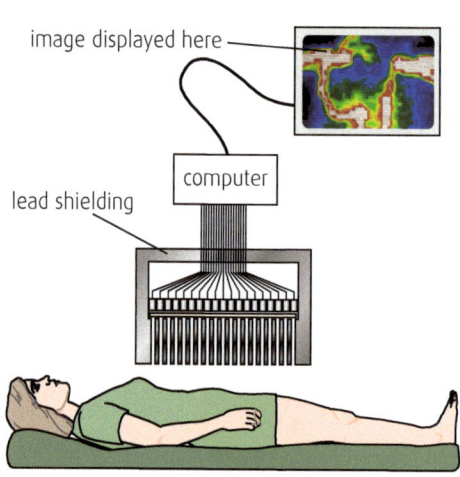

image displayed here
computer
lead shielding

Gamma camera.

Controlling the thickness of materials

When materials such as paper, floor coverings and aluminium foil are produced, it is important for the manufacturer to produce the material at a fixed thickness. If the material produced is too thick, then it would be too expensive and wasteful to produce. If the material is too thin, then it might break or snap when in use.

The aluminium foil passes through rollers that compress it to the correct thickness.

Beta radiation is detected from a source placed above the foil at the point where it leaves the rollers.

If the detector reading is too high, then this means that too little beta radiation has been absorbed by the foil because the foil is too thin. A computer is used to reduce the pressure of the rollers until the radiation is at the correct level for the normal thickness.

If the detector reading is too low, then this means that too much beta radiation has been absorbed by the foil because the foil is too thick. A computer is used to increase the pressure of the rollers until the radiation is at the correct level for the normal thickness.

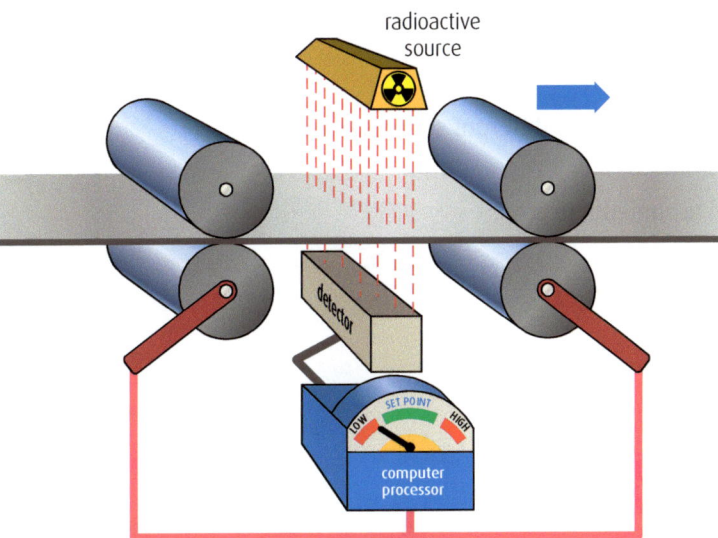

Radiation can be used to control the thickness of materials.

 ## TASK Quick questions

1. Describe one use of nuclear radiation to diagnose health problems.
2. Describe one use of nuclear radiation to treat health problems.
3. Describe one use of nuclear radiation in industry.

THINGS TO DO AND THINK ABOUT

Carry out research into more uses of radiation in medical and industrial applications. Investigate the uses of radioactive materials that are either naturally found on Earth or are made by humans. Construct a table for both that shows some of the uses of radiation.

THE NUCLEAR ENERGY DEBATE

Many radioactive materials occur naturally. For example, the fuel used in nuclear power stations is first obtained from uranium mines.

It then undergoes several processes before being used in fuel rods in nuclear power stations.

Nuclear power stations are used by many countries to produce electrical energy. They use radioactive materials as the source of nuclear energy.

In some countries, people and the government have decided to stop using nuclear power stations to generate electricity. The nuclear power stations in these countries will be dismantled and no further nuclear power stations will be constructed. These countries will have to use other sources of energy in power stations.

In contrast, some countries have decided that nuclear power stations will be used as part of a range of different sources of energy to generate electricity.

Uranium mining in Australia.

There are both arguments to support, and arguments to stop, the use of nuclear power stations.

ARGUMENTS AGAINST USING NUCLEAR FUEL AS AN ENERGY SOURCE

Accidents are dangerous

Accidents in nuclear power stations can cause very serious damage to people and the surrounding environment. Many people believe that the risk of serious accidents at nuclear power stations is not acceptable and that they should be dismantled.

Nuclear waste is dangerous

Used (or spent) nuclear fuel from power stations has to be sent to special recycling and storage facilities. Materials are sent by train, lorry and by sea. Many people think that the risk of accidents happening to materials while they are being transferred is not acceptable because any radioactive materials escaping during an accident could harm people, wildlife and the environment.

Nuclear waste is expensive to store

Spent nuclear fuel must be stored safely for many years before it reaches safer levels of radioactivity. This storage is very expensive to maintain and there are fears that, over time, radioactive materials could escape or leak into the environment causing damage, or even be stolen.

Torness nuclear power station.

Nuclear waste being transported by train to a storage facility.

ARGUMENTS FOR USING NUCLEAR FUEL AS AN ENERGY SOURCE

Little pollution produced from nuclear power

Nuclear power stations cause very little pollution of the atmosphere compared with power stations that burn fossil fuels.

Many people believe that the carbon dioxide gas released by burning fossil fuels in power stations is a major cause of climate change and global warming. It is said that global warming will cause irreversible damage to the Earth's climate if it continues and that emissions of carbon dioxide gas from fossil fuel power stations must be stopped by closing them. Some people also say that nuclear power stations are the only alternative that can provide enough electrical energy to meet the demands for electricity if the fossil fuel power stations are closed.

Spent nuclear fuel rods kept under water at a nuclear waste storage facility.

Needed to meet growing demand

Many people believe that nuclear power stations must be used to supply the growing demand for electrical energy.

In some countries, fossil fuel sources are running out. Some people believe that nuclear power stations are the only alternative to replace the huge amount of electrical energy currently supplied by fossil fuel power stations.

Many scientists believe that renewable energy alone could not meet the demands for electrical energy if both fossil fuel and nuclear power stations are phased out and so nuclear power stations should continue to be used.

Pollution from a coal-fired power station.

THINGS TO DO AND THINK ABOUT

Carry out more research into the debate about using nuclear energy to generate electricity.
Find information about recent accidents at nuclear power stations.
Find information about the predicted effects of global warming.
Write a summary of what you have found to compare the arguments for and against the use of nuclear energy.

 DON'T FORGET

There are arguments for and against the use of nuclear power stations.

EXTENDED QUESTIONS 1

WAVE CHARACTERISTICS

1. The following diagrams show waves being produced by vibrating a 'slinky spring'.

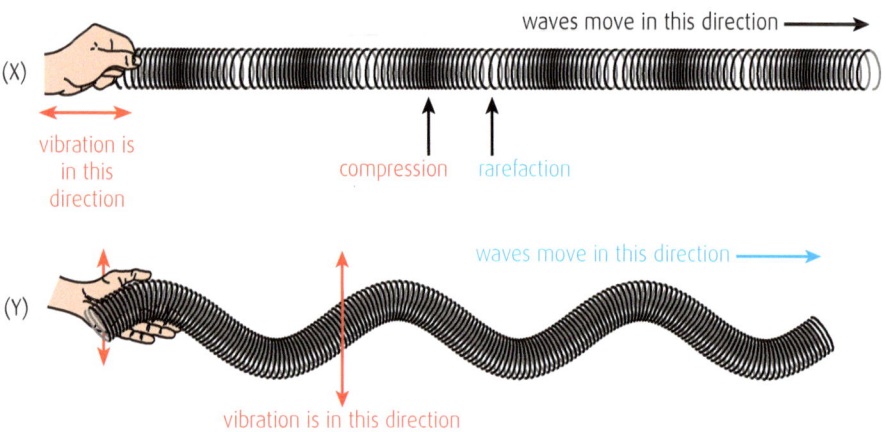

   Which waves are: (i) transverse; and (ii) longitudinal?

2. Determine the amplitude of the following waveform.

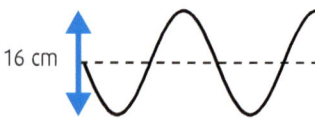

3. Determine the wavelength of the following waveform.

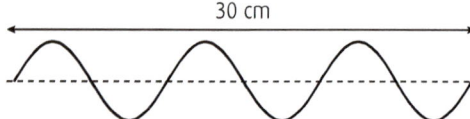

4. The following diagram represents a wave travelling from left to right.

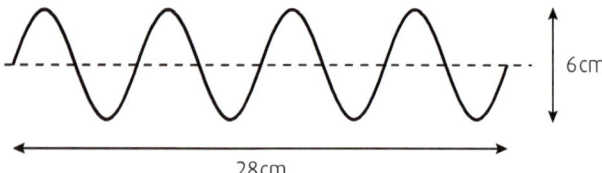

   Determine the wavelength and amplitude of the wave.

5. Waves are produced by a wave machine in a swimming pool at a frequency of 2 Hz. How many waves are produced in 50 seconds?

6. A person standing by the sea counts 12 waves passing a marker in 36 seconds. Calculate the frequency of the waves.

7. An oscilloscope displays an electronic signal.

   (a) What is the frequency of the signal?
   (b) What is the amplitude of the signal?

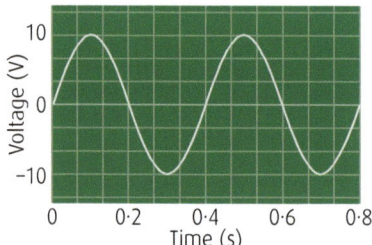

8. State which diagram shows waves that are:

(i) longitudinal; and (ii) transverse.

9. A wave pattern is shown in the diagram. The frequency of the waves is 5 Hz.

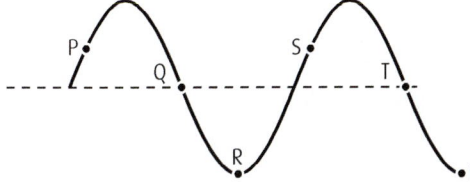

(a) Which two letters are a distance of one wavelength apart in this pattern?
(b) State what is meant by a frequency of 5 Hz.

10. Calculate the missing values in the table.

	Wave speed v (m s⁻¹)	Frequency f (Hz)	Wavelength λ (m)
(a)	1500	80	
(b)	340		1·7
(c)		950	3·2

Diagram 1: water waves at sea.

11. A tuning fork produces sounds with a wavelength of 0·4 m. Calculate the frequency of the waves in air if their speed is 340 m s⁻¹.

12. A wave generator used in a marine laboratory has a frequency of 12 Hz. The wavelength of the waves produced in a tank for one test was 0·3 m. Calculate the speed of the waves.

Diagram 2: a ringing bell producing sound waves.

13. Calculate the wavelength of sounds of frequency 650 Hz travelling through a steel railing (speed of sound in steel is 5200 m s⁻¹).

14. Water waves move 24 m in a time of 16 s. Calculate the speed of the waves.

15. Calculate how far water waves travel in 14 s when their speed is 4·2 m s⁻¹.

16. Calculate the time taken for water waves with a speed of 5·8 m s⁻¹ to travel a distance of 26·1 m.

17. Waves moving on the surface of a canal take 16 s to travel a distance of 4·2 m.
(a) Calculate the speed of the waves.

A student standing on the canal bank counts 18 waves passing a point in 24 s.
(b) Calculate the frequency of the waves.
(c) Calculate the wavelength of the waves.

18. Students are observing water waves produced in a tank at a marine laboratory.

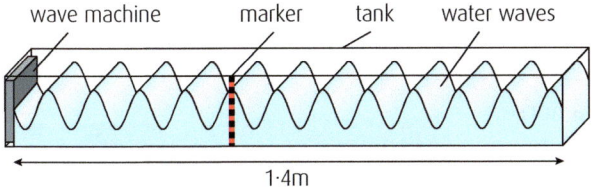

One student measures 5.6 s for one wave to travel the length of the tank.
(a) Calculate the speed of the waves.

Another student counts 28 waves passing a marker on the tank in 22·4 s.
(b) Calculate the frequency of the waves.
(c) Calculate the wavelength of the waves.

EXTENDED QUESTIONS 2

SOUND

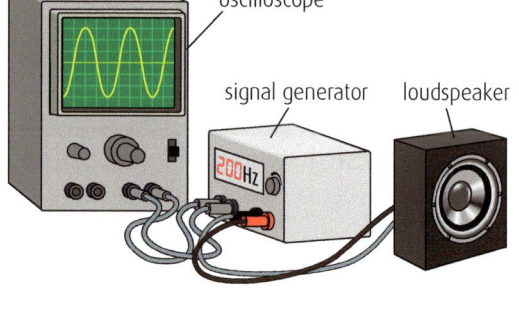

oscilloscope

signal generator   loudspeaker

1. A signal generator is connected to an oscilloscope and a loudspeaker to investigate sound waves.

   The signal generator is switched on and adjusted to produce a sound from the loudspeaker. The oscilloscope trace is shown in Figure 1.

   (a) Copy and complete Figure 2 to show the trace obtained when the loudness of the sound is reduced.

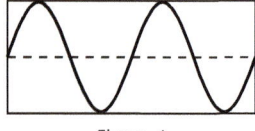 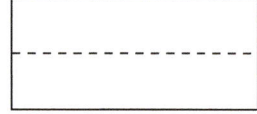

Figure 1   Figure 2

   (b) Copy and complete the following sentence:
   The [____] control is used to adjust the number of waves produced each second.

2. A student investigates the speed of sound in water.
   The student sets up the experiment on the right:
   The student strikes the metal plate with the hammer. The electronic timer measures the time taken for the sound to travel from microphone 1 to microphone 2.
   The time recorded on the electronic timer is 0·0008 s.
   (a) Calculate the speed of sound in water.
   (b) Suggest one improvement that could be made to the experiment to improve the accuracy of the calculated value.

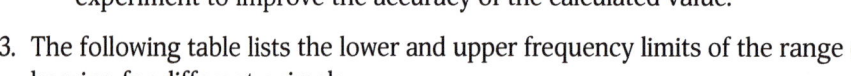

electronic timer

microphone 1   water   microphone 2

1·16m

3. The following table lists the lower and upper frequency limits of the range of hearing for different animals.

Animal	Frequency of lower limit of hearing (Hz)	Frequency of upper limit of hearing (Hz)
Mouse	450	65 000
Cat	50	65 000
Dog	65	45 000
Horse	60	34 000
Human	20	20 000
Elephant	15	12 000

   (a) What is the lowest frequency that can be heard by a horse?
   (b) Which animal in this table can hear the narrowest range of frequencies?
   (c) A dog whistle is used to alert a dog that its owner is calling. The frequency of the whistle is above the human hearing range. A dog owner wants to summon a dog that is running in a field close to horses, but the owner does not want to annoy the horses.
   Suggest a possible operating frequency for the whistle.
   (d) The speed of sound in air is 340 m s^{-1}.
   Calculate the wavelength of the lowest frequency of sound heard by a cat.

ELECTROMAGNETIC SPECTRUM

1. State one hazard associated with ultraviolet radiation.

2. Name a detector of: (a) X-rays; (b) infra-red radiation.

NUCLEAR RADIATION

1. Write down the names of the particles in an atom and the charge carried by each of the particles.

2. Write down the names of the particles labelled in the diagram of the structure of an atom.

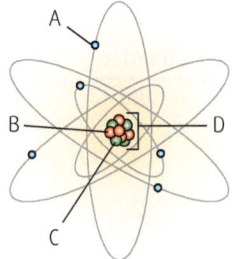

3. Name one artificial and one natural source of radiation.

4. (a) Write down the names and symbols for the three types of nuclear radiation.
   (b) State what each type of radiation consists of.

5. For the each of the three types of nuclear radiation, explain how far they can travel in air, and which materials can absorb them.

6. Name one detector for nuclear radiation.

7. A radioactive material emits two types of radiation, labelled X and Y. The diagram shows how far these two types of radiation can travel through different materials.
   (a) Identify radiation X.
   (b) Identify radiation Y.

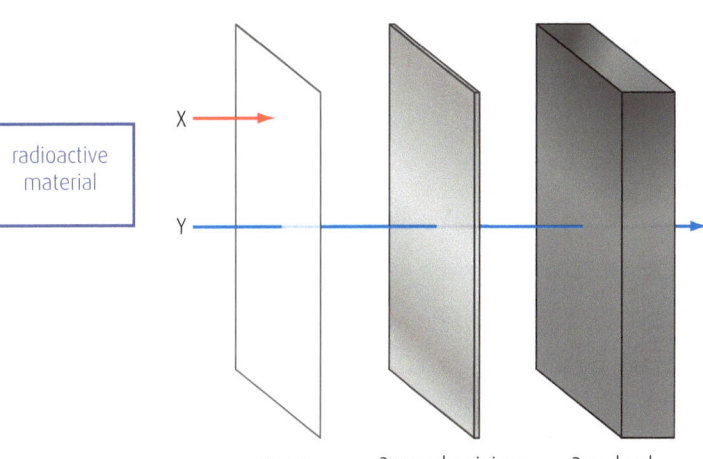

8. Explain what happens to the atoms of a radioactive element when it decays.

9. Write a definition of the activity of a radioactive substance.

10. State what is meant by the term 'half-life'.

11. Different types of radiation in the electromagnetic spectrum can be used in medicine for the diagnosis and the treatment of many illnesses. The table gives information about some types of radiation used in medicine and industry.

Type of radiation	Application in medicine	Application in industry	Possible hazard
X-rays	Imaging broken bones	Scanning cargo at airports	A
B	Treating certain skin conditions	Detecting forged banknotes	Can damage eyesight
Infra-red	C	Detecting wasted heat energy from buildings	Causes skin burns
Microwave	Destroying tumours	D	Can cause internal heating of body tissues

   (a) State a possible hazard for A.
   (b) State the name of radiation B.
   (c) State one application in medicine for C.
   (d) State one application in industry for D.

12. (a) Write a list of the arguments for the use of nuclear power stations.
    (b) Write a list of the arguments against the use of nuclear power stations.

DYNAMICS AND SPACE

SPEED AND ACCELERATION 1

AVERAGE AND INSTANTANEOUS SPEED

The speed v of a moving object is the distance d that the object moves divided by the time taken t.

$$\text{speed} = \frac{\text{distance}}{\text{time}} \qquad v = \frac{d}{t}$$

Average speed

Some speed cameras monitor the average speed of cars.

The **average speed** of a moving object such as a car is calculated using two measurements and substituting these two measurements into the formula for average speed.

An example of these two measurements are:

- the distance d between two lamp-posts (or any two convenient markers);
- the time taken t to travel this distance (measured with a stopwatch).

The average speed is calculated using: $\bar{v} = \dfrac{d}{t}$

The line (or bar) above the v is a shorthand way of indicating the average speed.

The relationship is sometimes written as: $d = \bar{v}\, t$

Rearranging this formula to calculate t gives: $t = \dfrac{d}{\bar{v}}$

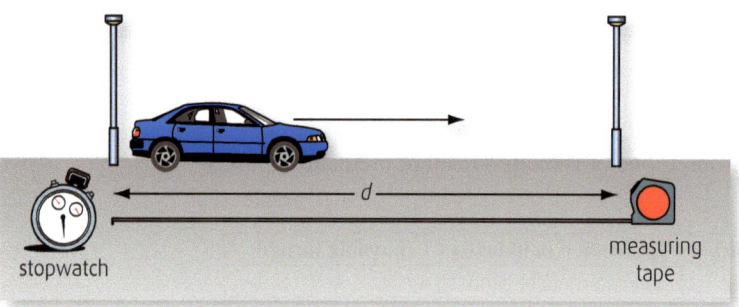

stopwatch

measuring tape

Calculating the average speed of a car.

JUST A WEE NOTE

There is no penalty if you don't include the bar above the v.

EXAMPLE

A snail travels 0·55 m in 1100 s. Calculate the average speed of the snail.

ANSWER

$$\bar{v} = \frac{d}{t} = \frac{0.55}{1100} = 0.0005 \text{ m s}^{-1}$$

EXAMPLE

A car travels at an average speed of 11·2 m s^{-1} for 125 s. Calculate the distance travelled by the car.

ANSWER

$$d = \bar{v}\, t = 11.2 \times 125 = 1400 \text{ m (1·4 km)}$$

EXAMPLE

EXAMPLE

A school bus travels from a village towards a school. It travels 726 m to the first bus stop in 80 s. It waits at this bus stop for 45 s. It then travels 1300 m to the next bus stop in a time of 85 s. It waits at this bus stop for 65 s. It completes the final 1207 m in a time o f 75 s to reach the school.

Calculate the average speed of the bus during this journey.

ANSWER

Total distance travelled by the bus
= 726 + 1300 + 1207= 3233 m
Total time for the journey
= 80 + 45 + 85 + 65 + 75 = 350 s

Note that the time for the average speed is the total time – this includes time when the vehicle is stationary.

$$\bar{v} = \frac{d}{t} = \frac{3233}{350} = 9.2 \text{ m s}^{-1}$$

Some speed cameras monitor the instantaneous speed of cars.

Instantaneous speed

The instantaneous speed of a moving object is the average speed over a small time interval (usually much shorter than one second).

A light gate is often used to measure small intervals of time.

The light gate is connected to an electronic timer, which starts when a light beam is interrupted by the card and stops when the beam resumes.

The **instantaneous** speed of the trolley as it passes through the light gate is: $v = \dfrac{\text{card width}}{\text{interrupted time}}$

The **average** speed of the trolley as it passes through the light gate is:

$$\bar{v} = \frac{\text{total lenght of slope}}{\text{time taken by trolley to travel down slope}}$$

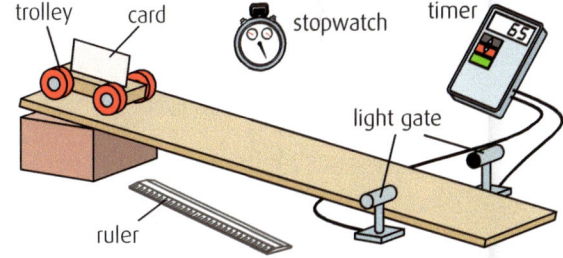

trolley   card   stopwatch   timer

light gate

ruler

EXAMPLE

On a motorway, a car passed a speed camera that recorded the car travelling 3 m in 0·08 s. Calculate the instantaneous speed of the car.

ANSWER

$$v = \frac{d}{t} = \frac{3}{0.08} = 37.5 \text{ m s}^{-1}.$$

EXAMPLE

During a 100 m sprint by Usain Bolt, a video camera was used to film and record his running time. During one section of the race, Usain ran a distance of 0·015 m in a time interval of 0·0012 s. Calculate his instantaneous speed for this section of the race.

ANSWER

$$v = \frac{d}{t} = \frac{0.015}{0.0012} = 12.5 \text{ m s}^{-1}.$$

JUST A WEE NOTE

A common mistake is using the complete length of a race when calculating instantaneous speed.

THINGS TO DO AND THINK ABOUT

1. A cross-country runner travels 4500 m in a time of 750·0 s. Calculate the runner's average speed.

2. Calculate the time taken by a high-speed train to travel 252 000 m at an average speed of 100 m s⁻¹. Give the answer in both seconds and in hours.

3. Calculate the distance travelled by an aeroplane in two hours flying at an average speed of 134 m s⁻¹.

4. Calculate the instantaneous speed of a car that travels 2·4 m in 0·08 s.

5. Explain the difference between average speed and instantaneous speed.

DON'T FORGET

When both the card width and the slope length are given, great care must be taken to use the correct substitution for d in the formula $v = \dfrac{d}{t}$.

SPEED AND ACCELERATION 2

SPEED-TIME GRAPHS

Speed–time graphs are used to display the motion of objects that are speeding up, slowing down or travelling at a constant speed.

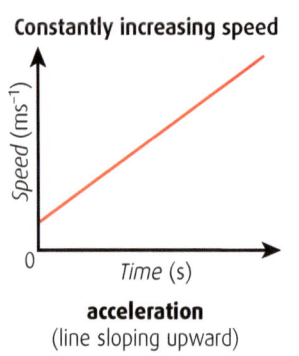

Constantly increasing speed

acceleration
(line sloping upward)

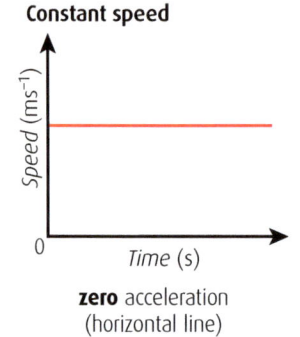

Constant speed

zero acceleration
(horizontal line)

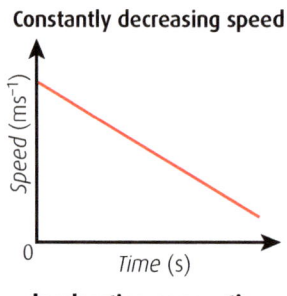

Constantly decreasing speed

deceleration or **negative**
acceleration
(line sloping downward)

For acceleration, the steeper the slope (greater angle), the greater the acceleration:

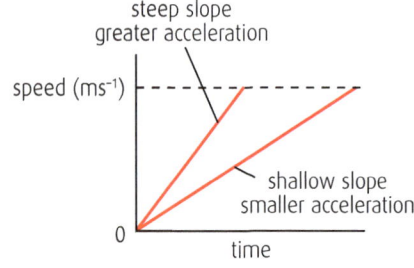

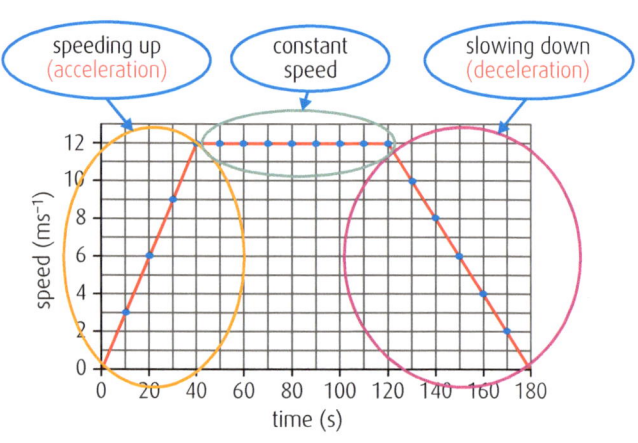

> **EXAMPLE**
>
> The speed of a car was measured at different times by recording the readings on its speedometer every 10 s.
>
> The results are shown in the table.

Time (s)	0	10	20	30	40	50	60	70	80	90	100	120	130	140	150	160	170	180
Speed (m s-1)	0	3	6	9	12	12	12	12	12	12	12	12	10	8	6	4	2	0

These readings can be plotted on a speed–time graph. The different sections of the graph indicate the different parts of the car's journey.

The car's speed–time graph shows that:

- the car accelerated (for 40 s)

- then travelled at a constant speed (for 80 s)

- then decelerated (for 60 s).

EXAMPLE

The following graph shows the speed–time plot of a bus leaving a bus stop and travelling to the next stop.

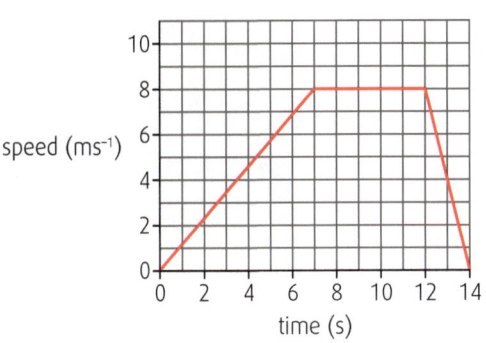

Describe the motion of the bus.

ANSWER

The graph shows that:
- the bus accelerated from zero to 8 m s^{-1} in 7 s
- then continued at a constant speed of 8 m s^{-1} for a further 5 s
- then decelerated from 8 m s^{-1} to zero in 2 s.

EXAMPLE

The speed–time graph of a bat approaching an insect in flight is shown. Describe the motion of the bat.

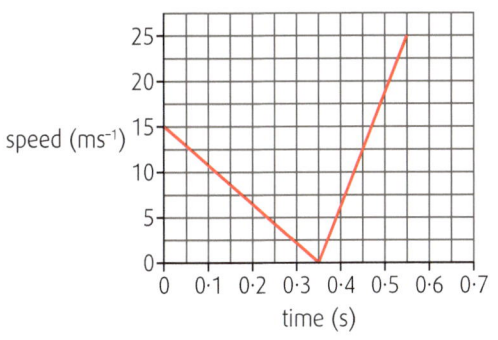

DON'T FORGET

Speed–time graphs show constant speed or acceleration or deceleration.

ANSWER

The bat starts at a speed of 15 m s^{-1} and decelerates for 0·35 s until zero speed. The bat then accelerates from zero for 0·2 s to a speed of 25 m s^{-1}.

THINGS TO DO AND THINK ABOUT

1. Speed–time graphs for different types of motion are shown.

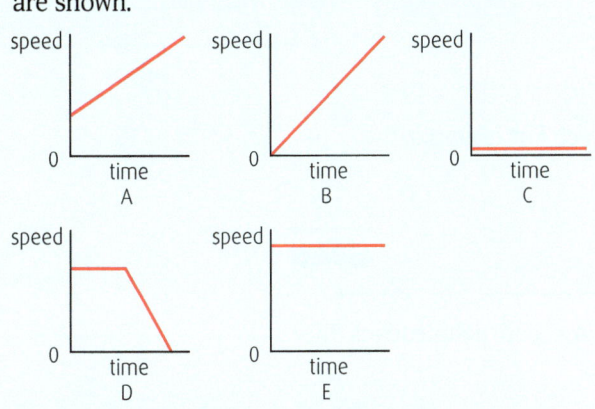

Which graph could represent a car:
(a) Travelling at a constant speed on a fast road?
(b) Driving along a road at constant speed then braking?
(c) Driving away from traffic lights?
(d) Accelerating to overtake another vehicle?
(e) Reversing carefully along a drive?

2. The following graph represents the movement of a delivery van moving from one customer to the next.

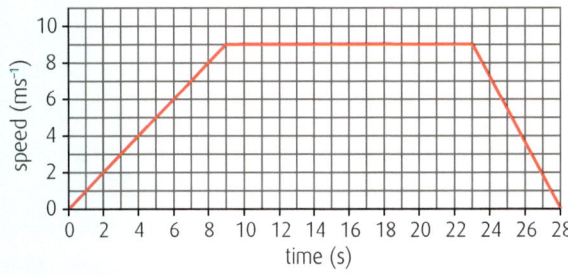

(a) State the time for which the van accelerated.
(b) State the time for which the van travelled at constant speed.
(c) State the time for which the van decelerated.

SPEED AND ACCELERATION 3

SPEED-TIME GRAPHS AND DISTANCE TRAVELLED

The speed–time graph of an object can be used to calculate how far the object moves.

The area under the line in the graph is equal to the distance.

EXAMPLE

The speed–time graph of a train for the first 10 s after it leaves the station platform is shown in the graph. Calculate the distance travelled by the train in this time.

Distance = area under graph = area of triangle.
= ½ base × height = ½ × b × h

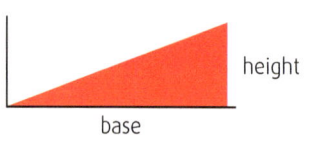

Use the values from the speed and time axis:
base = 10 s, height = 30 m s⁻¹
So distance d = area under graph = ½bh
d = ½ × 10 × 30 = 150 m.

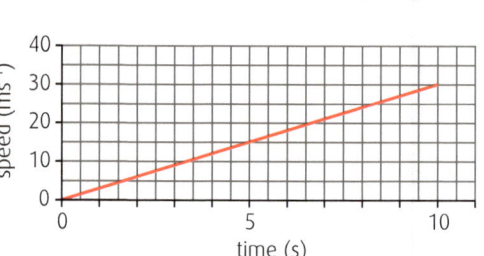

EXAMPLE

The speed–time graph of a bus travelling from one bus stop to the next is shown. Calculate the distance travelled by the bus during this time.

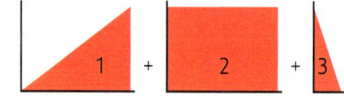

Distance = area under graph
= area of triangle + area of rectangle + area of triangle.

To get the whole distance, add up the areas for each shape of the graph.

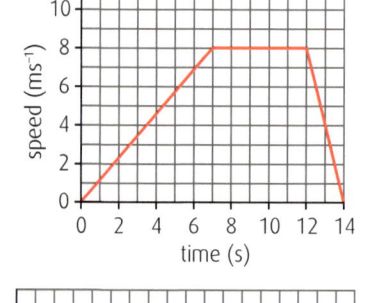

From the graph:
- area 1 (triangle): base = 7, height = 8; area = ½bh = ½ × 7 × 8 = **28**
- area 2 (rectangle): base = 5, height = 8; area = bh = 5 × 8 = **40**
- area 3 (triangle): base = 2, height = 8; area = ½bh = ½ × 2 × 8 = **8**

Distance = total area under graph = 28 + 40 + 8 = 76 m.

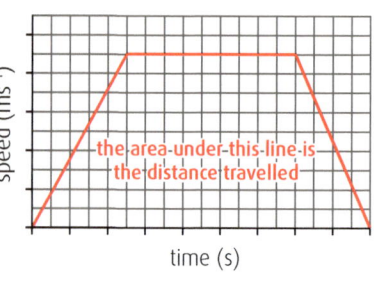

ACCELERATION

Acceleration is the rate of change of speed of a moving object. For example, the diagram shows a car speeding up from rest.

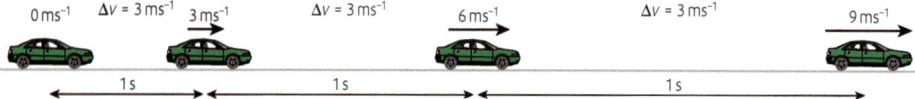

The change in speed is denoted by Δv (the triangle symbol Δ called 'delta' means 'the change in', so Δv means the change in speed).

After each second, the speed of the car is greater. In this case, it is always greater by the same amount of 3 m s⁻¹.

When the change in speed is always the same amount, this is called constant or uniform acceleration.

In this example, the speed of the car changes by 3 m s⁻¹ **every** second.

This can be written out as three metres **per** second **per** second (sometimes called metres per second squared).

The unit for acceleration is **m s^{-2}**.

Acceleration can be determined by calculating:

$$\frac{\text{change in speed}}{\text{time taken for change}} \text{ or } a = \frac{\Delta v}{t}$$

where a = acceleration (m s^{-2}), Δv = change in speed (m s^{-1}) and t = time taken for change (s).

Rearranged versions of the formula: $\Delta v = at$ and $t = \dfrac{\Delta v}{a}$

EXAMPLE

When overtaking a bus, a car changes its speed by 9 m s^{-1} in 4 s.

Calculate the acceleration of the car.

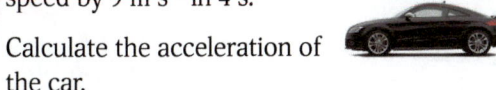

ANSWER

$$a = \frac{\Delta v}{t} = \frac{9}{4} = 2.25 \text{ m s}^{-2}$$

Experiment to measure acceleration

There are different methods that can be used to measure acceleration. One method uses a light gate connected to a computer.

A trolley has two identical cards attached to it. The width of the cards is entered into the computer. The trolley is released from the top of the slope and each card, in turn, passes through a light gate at the foot of the slope.

The computer measures the time taken for each card to pass through the light gate and the total time for the change in speed. The computer calculates the change in speed and then the acceleration.

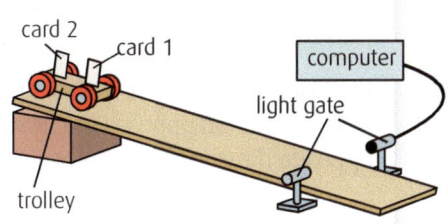

Measuring acceleration.

Sample results:

Run no.	Δv or change in speed (m s^{-1})	t or time taken for change of speed (s)	Calculate: $a = \dfrac{\Delta v}{t}$ (m s^{-2})
1	0·112	0·200	0·56
2	0·135	0·250	0·54
3	0·121	0·220	0·55
average $a = \dfrac{a_1 + a_2 + a_3}{3} =$	$\dfrac{0.56 + 0.54 + 0.55}{3}$	$= 0.55$ m s^{-2}	

TASK Quick question

The following graph represents the motion of a firework during the first 0·2 s after its launch.

Calculate the distance travelled during these 0·2 s.

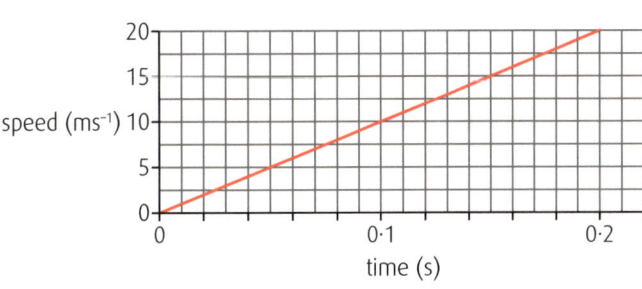

THINGS TO DO AND THINK ABOUT

1. Copy and complete the table.

	Δv (m s^{-1})	t (s)	a (m s^{-2})
(a)	14	0·7	
(b)	0·05		0·02
(c)		1·4	270

RELATIONSHIP BETWEEN FORCES, MOTION AND ENERGY 1

FORCES

A force applied to an object can change the object's **shape**, **speed** and **direction** of travel, such as pulling a length of Plasticine, pushing a sledge or striking a tennis ball.

The size of a force is measured in newtons (N) and can be measured using a newton balance.

Force of friction

Friction is a force that is present whenever an object moves or tries to move. The force of friction opposes the motion of an object. This means that it acts in the opposite direction to the movement of the object.

For example, **air resistance**, or drag, is a frictional force and acts in the direction **opposite** to the motion.

Friction can be decreased by streamlining the shape of a vehicle's bodywork so the air passes over more smoothly.

This will improve the fuel consumption of the lorry. The friction on a vehicle is deliberately increased when the lorry brakes. Brake pads grip part of the wheel during braking, creating friction.

Examples of forces.

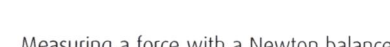

Measuring a force with a Newton balance.

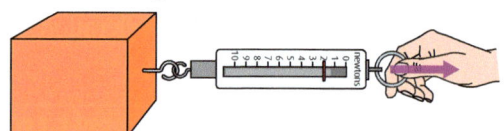

air resistance

direction of travel

Air resistance acting in the opposite direction to motion.

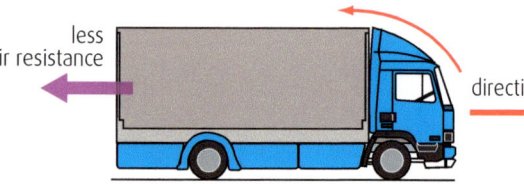

less air resistance

direction of travel

Reducing air resistance by streamlining the shape of a lorry.

NEWTON'S FIRST LAW: BALANCED FORCES

When the forces on an object are **equal** and **opposite**, we say that the forces are **balanced**.

An example of balanced forces is a tug-of-war. Both teams pull on the rope with equal force, but in opposite directions.

Balanced forces in tug-of-war.

Newton's First Law states that, when the forces acting on an object are balanced, if the object is at rest, it will remain at rest.

Newton's First Law also states that, when the forces acting on a moving object are balanced, it will move at a constant speed.

DON'T FORGET

Balanced forces mean either no motion, or a constant speed.

Examples of balanced forces

EXAMPLE

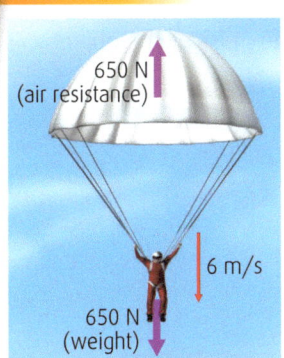

A chair resting on the floor.

The downward force (weight) of the chair is balanced by an upward (reaction) force from the floor acting on the chair legs.

EXAMPLE

If the forward force from the cyclist is equal to the frictional force in the opposite direction, then the cyclist will move at a constant speed.

The converse is also true: when the speed of a moving object is constant, then the forces acting on it are balanced.

EXAMPLE

650 N (air resistance)

6 m/s

650 N (weight)

A parachutist descends at a constant speed of 6 m s^{-1}. The parachutist's weight is 650 N and this force acts vertically downwards. As the speed is constant, this force must be the same as the air resistance (frictional force) acting upwards.

EXAMPLE

The forces are balanced on the car moving at a constant speed. In the horizontal direction, the forward force supplied by the car engine is balanced by the opposing frictional force of air resistance.

In the vertical direction, the downward force of the weight of the car is balanced by the upward reaction force exerted by the ground on the car tyres.

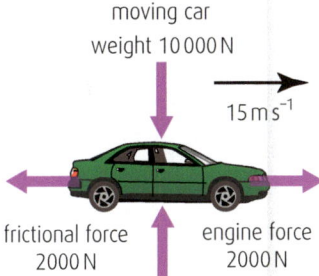

moving car weight 10 000 N

15 m s^{-1}

frictional force 2000 N

engine force 2000 N

reaction force 10 000 N

EXAMPLE

If a spacecraft is in outer space, then it is far away from the effect of gravity from planets or stars and no gravitational force acts on it.

This is an unusual case because, in space, the spacecraft can use its engines to accelerate to a very high speed because there is no air resistance. If the engines are switched off, there is no frictional force to oppose its movement and slow it down (because there is no air in space).

This means that there is **no force** acting on the spacecraft, so it will move at a constant speed.

THINGS TO DO AND THINK ABOUT

1. State three things that a force can do to an object.
2. If the forces acting on an object are equal and opposite, how are they described?
3. State what happens to a stationary object when the forces acting on it are balanced.
4. State what happens to a moving object when the forces acting on it are balanced.

RELATIONSHIP BETWEEN FORCES, MOTION AND ENERGY 2

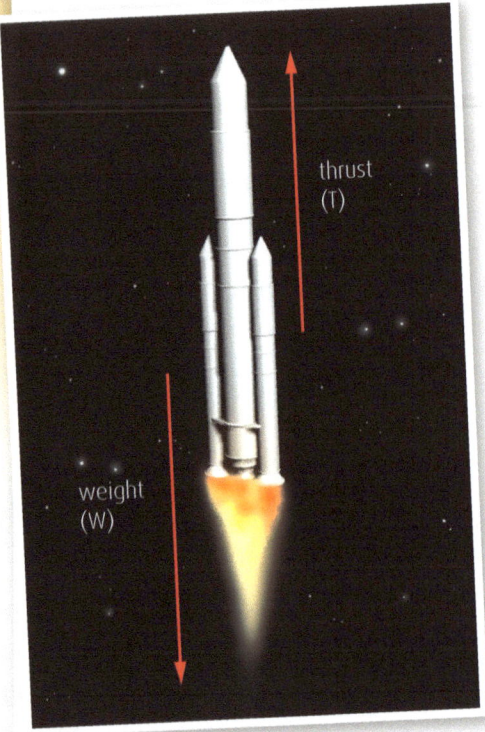

thrust
(T)

weight
(W)

NEWTON'S SECOND LAW: UNBALANCED FORCES

When a rocket is launched, the force acting on it upwards (called the thrust) is **greater** than the force acting on it downwards (the weight of the rocket).

This means that the forces acting on the rocket are unbalanced. The unbalanced force causes the rocket to accelerate upwards.

When the forces acting on an object are **unbalanced**, the object no longer moves at a constant speed, but will **accelerate**.

When an unbalanced force acts on an object, the **size** of the acceleration *a* depends on the **size** of the unbalanced force (greater unbalanced force = greater *a*) and the mass of the object (greater mass = smaller *a*).

Small force, smaller acceleration

Larger force, greater acceleration

Newton's Second Law combines these facts:

$F = ma$ 　　　　　where F = unbalanced force in newtons (N), m = mass of object in kilograms (kg) and a = acceleration of object (m s^{-2}).

Rearranged versions of the formula are:

$$a = \frac{F}{m}$$

$$m = \frac{F}{a}$$

Greater mass, smaller acceleration

Small mass, larger acceleration

One type of re-entry for spacecraft is known as 'ballistic descent' (this is the method used by the Soyuz spacecraft). The space vehicle is steered into the atmosphere to return, almost in freefall, directly to the surface. Ballistic re-entry is fast and typically takes less than one hour. Once the descent has started, limited control can be exerted on the craft, apart from deploying the parachutes when it is close to the Earth's surface. The descent module is designed to have a specific shape to protect the interior from overheating.

The disadvantage of this method is that a great deal of equipment has to be discarded in space, leaving only the module to make the descent. The size of the descent module is also limited. Much of the mass of the returning descent module consists of its heat shield, which protects the occupants and cargo from the heat produced.

HEAT SHIELD DESIGN

A spacecraft's heat shield has several features to help reduce the heating of the interior by insulating the occupants from the extreme temperatures produced by re-entry.

Dissipation

Some heat shields use a process called **dissipation**. The heat energy is absorbed by insulating tiles that cover the spacecraft. Some of the absorbed heat energy is re-radiated from the spacecraft back into the atmosphere to maintain an acceptable temperature level inside the astronaut's compartment. The Space Shuttle used this technique.

Ablation

Some heat shields use a complex process known as **ablation**.

One ablation process involves the disintegration of the surface material of the heat shield at extreme temperatures, removing heat energy as the material leaves the shield. The temperature of the craft is thus reduced.

Another ablation process occurs at high temperatures when some of the material inside the heat shield is changed into a gas by the heat energy. The pressure of the gas forces it out of the heat shield, removing more heat energy.

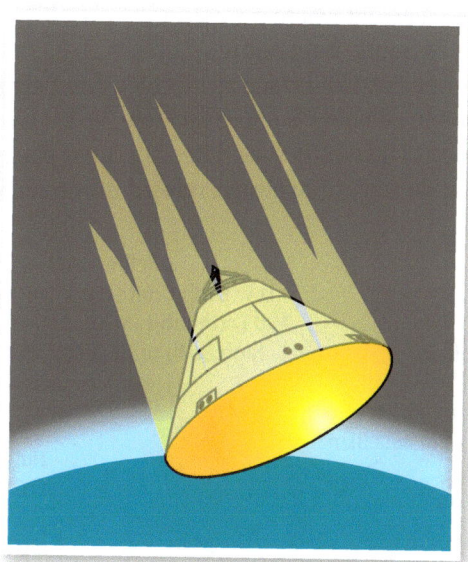

Some heat shields use ablation to remove heat on re-entry.

> **DON'T FORGET**
>
> The technology needed for space exploration is very complex. Safety is always a priority.

THINGS TO DO AND THINK ABOUT

1. Find out about the Apollo 13 mission to the Moon. Write a summary of what went wrong during the mission and how the crew were able to make repairs to allow them to return safely to Earth.

2. Carry out research into manned and unmanned space missions. Construct a table and indicate the purpose of each mission and whether it was successful.

SATELLITES 1

Satellites are placed into orbit around the Earth and are used for many different purposes.

Satellites are used for communication, to gather data about the Earth (e.g. for weather forecasting), for military observations and information-gathering, and even for the observation of electromagnetic radiation from space.

HEIGHT OF ORBIT

The height of a satellite's orbit above the Earth's surface depends on its intended use. Satellites in low orbits must travel faster than satellites in higher orbits. The International Space Station orbits the Earth once every 90 minutes and thus orbits the Earth 16 times each day.

The time taken for a satellite to go once round the Earth depends on the height of the orbit above the Earth. The **period** of the satellite is the time taken for it to complete one rotation around the Earth. The further away the satellite is from the Earth, the longer the period.

Geostationary satellites

A **geostationary satellite** orbits at a height where a complete orbit of the Earth takes 24 hours. It therefore remains above the same point on the Earth's surface as it orbits. The height that allows a satellite to make one orbit in 24 hours is approximately 36 000 km above the Earth's surface. Communications and weather satellites are often placed in geostationary orbits because their data are then always related to the same area on the Earth's surface.

R_1 (radius of satellite 1)
R_2 (radius of satellite 2)

satellite 1
(short period of orbit –
closer to the Earth's surface)

Earth

R_1

R_2

satellite 2
(long period of orbit –
further from the Earth's surface)

The further away from the Earth,
the longer the period of the satellite

The time taken to orbit the Earth depends on the
height of the satellite.

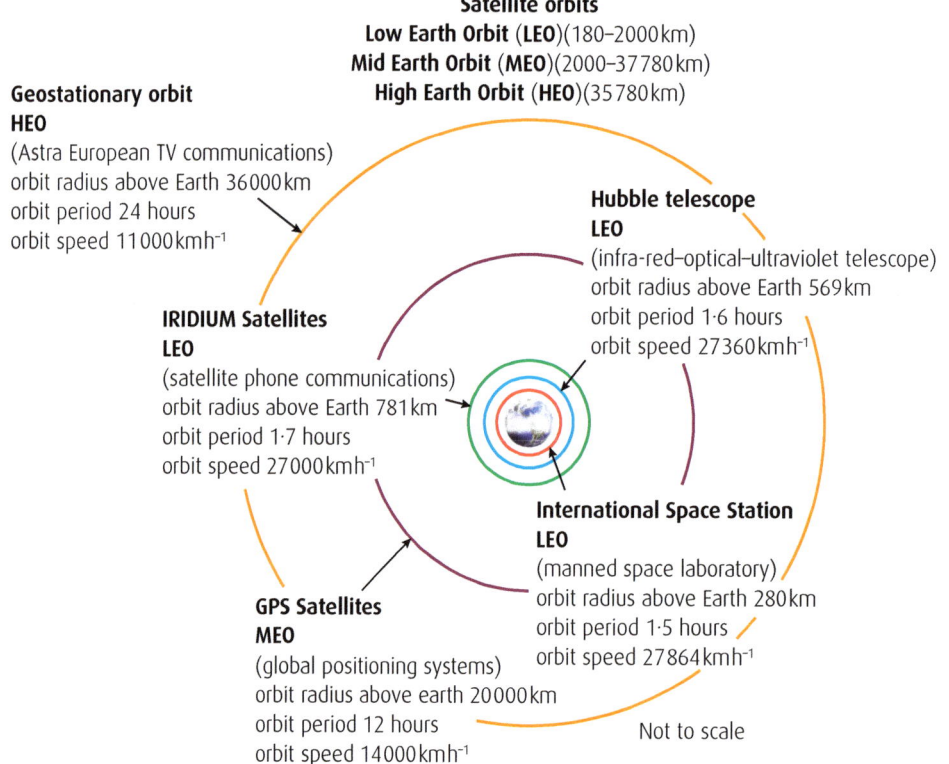

Satellite orbits
Low Earth Orbit (**LEO**)(180–2000km)
Mid Earth Orbit (**MEO**)(2000–37780km)
High Earth Orbit (**HEO**)(35780km)

Geostationary orbit
HEO
(Astra European TV communications)
orbit radius above Earth 36000km
orbit period 24 hours
orbit speed 11000kmh⁻¹

Hubble telescope
LEO
(infra-red–optical–ultraviolet telescope)
orbit radius above Earth 569km
orbit period 1·6 hours
orbit speed 27360kmh⁻¹

IRIDIUM Satellites
LEO
(satellite phone communications)
orbit radius above Earth 781km
orbit period 1·7 hours
orbit speed 27000kmh⁻¹

International Space Station
LEO
(manned space laboratory)
orbit radius above Earth 280km
orbit period 1·5 hours
orbit speed 27864kmh⁻¹

GPS Satellites
MEO
(global positioning systems)
orbit radius above earth 20000km
orbit period 12 hours
orbit speed 14000kmh⁻¹

Not to scale

Different types of orbits used by satellites.

USE OF PARABOLIC REFLECTORS TO SEND AND RECEIVE SIGNALS

Satellite communication is a two-way process. Radio waves sent from a ground station transmitter are used to carry control signals to a satellite. The satellite has a receiver that detects and processes the signals from Earth. Radio waves sent from a transmitter on the satellite are used to carry information, telephone signals and data from on-board sensors back to a receiver on Earth.

Curved reflectors on aerials or receivers can make the received signal stronger. Curved reflectors are often used in telecommunication systems such as satellite TV, TV link boosters (across the country), repeaters or in satellite communication.

A **dish aerial** can be used either to **receive** or to **transmit** a radio (or microwave) signal.

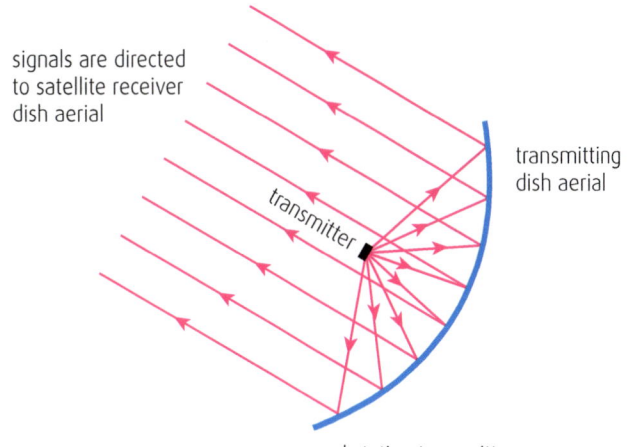

Signals are transmitted to satellites from ground stations.

The transmitter is placed at the focus of the curved reflector. The waves from the transmitter are then reflected by the dish aerial to produce a parallel beam of waves that is stronger than a beam emitted and spread in all directions. This beam is directed to the receiver dish aerial of the satellite. A dish with a larger diameter reflects more waves, so the signal detected at the receiver will have more energy.

The receiver is placed at the focus of the curved reflector. This concentrates the signal, which improves the reception. This is how satellite television is received in our homes.

Signals are received by aerial dishes on satellites.

THINGS TO DO AND THINK ABOUT

Carry out research into satellites that are in LEO, MEO and HEO orbits.

Group together satellites that have common uses – for example, weather forecasting satellites, communication satellites and global positioning satellites.

SATELLITES 2

INTERCONTINENTAL COMMUNICATION

Radio waves cannot travel directly to distant receivers on Earth because of the curvature of the Earth's surface. The signals are transmitted to a satellite, which then retransmits the signal to a receiver on the ground.

COMMUNICATION WITH SATELLITES

The radio signals used to communicate with satellites travel at the speed of light (3×10^8 m s^{-1}). Satellites are a significant distance from the Earth's surface. This means that there is a time delay between sending and receiving a signal.

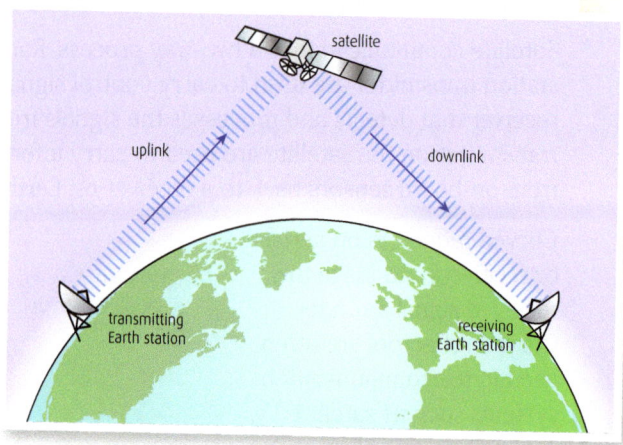

Satellites are used to transmit signals over the curved surface of the Earth.

EXAMPLE

A global positioning system communications satellite is in orbit 20 000 km above the Earth.

Calculate the time taken for a signal sent from Earth to reach the satellite.

20 000 km = 20 000 × 1000 m = 2×10^7 m

$$t = \frac{d}{v} = \frac{2 \times 10^7}{3 \times 10^8} = 0{\cdot}067 \text{ s}$$

A signal sent from a navigation satellite takes 0·07 s to reach a receiver on Earth. Calculate the distance of the satellite from the receiver.

$$d = vt = 3 \times 10^8 \times 0{\cdot}07 = 2{\cdot}1 \times 10^7 \text{ m (21 000 km)}$$

RANGE OF APPLICATIONS OF SATELLITES

Applications satellites have had an overwhelming effect on our day-to-day lives. These are satellites that have been placed in orbit for a specific purpose.

These satellites are used for:

- worldwide communication and data transfer – this has improved communications and has had an impact on our everyday lives;
- global and local weather information and forecasting, including tracking of hurricanes and other storm systems;
- monitoring of long-term climate change – this has highlighted the importance of reducing activities that lead to global warming;
- pollution monitoring – for example, some areas are exposed to increased ultraviolet radiation as a result of the deterioration of the ozone layer;
- provision of global positioning systems – satellite navigation is now in common use;
- military observations;
- commercial entertainment – satellite broadcasting means that events can be viewed worldwide almost as soon as they happen.

Satellites have had a huge impact on our everyday lives. For instance, weather forecasting has become very accurate as a result of using instant data from satellites. This has helped farmers to determine the best times to plant and harvest crops.

Numerous medical advances have developed from space research.

- Space research led to the development of devices to measure the amount of infra-red radiation emitted from distant stars and planets. This expertise was used to produce thermometers to determine body temperature accurately without direct contact with a patient by measuring infra-red radiation emitted from the inner ear. This has greatly reduced the incidence of cross-infection in hospitals.

- Water purification devices designed for use during space travel have been adapted to help patients with kidney disease.

- Technology developed to improve images of the Moon has been adapted for use with body imaging and scanning devices (e.g. in computer-aided tomography and magnetic resonance imaging).

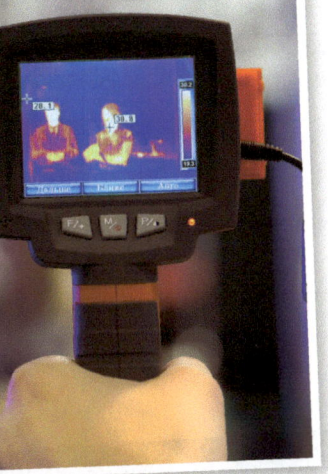

Thermometer that measures infrared radiation emitted from the body.

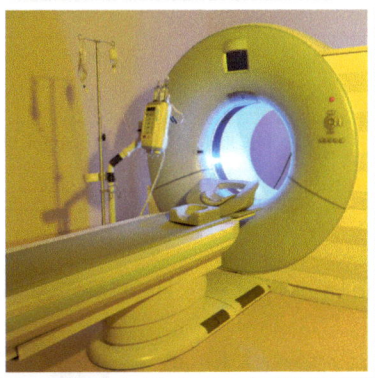

MRI scanners developed from technology that was used to observe the moon.

Uses of satellites.

Developments in space technology have provided benefits to medicine.

 THINGS TO DO AND THINK ABOUT

Carry out research to identify a satellite used to capture data for weather forecasting. Write a short summary of: (a) which part of the Earth it orbits; and (b) what type of data it gathers. For example, a satellite might contain infra-red sensors to determine the temperature of different parts of the Earth's surface.

 **DON'T FORGET**

Satellites are used to gather useful data and send information back to Earth.

COSMOLOGY

Our universe contains many different objects. As new information from space exploration is received, the list of the objects that have been discovered, or are thought to theoretically exist, continues to grow.

ASTRONOMICAL TERMS

Some of the astronomical terms used to describe objects in our universe are given in the table.

Term	Definition	Facts
Planet	A natural satellite that orbits a star	Eight planets orbit our local star (the Sun). Note: Pluto is now described as a dwarf planet
Exoplanet	A planet outside our solar system, orbiting another star	To date, more than 1800 exoplanets have been discovered and this number is increasing all the time
Moon	A natural satellite that orbits a planet	Astronomers have discovered at least 140 moons orbiting other planets in our solar system
Star	A mass of plasma that emits heat and light	Vast amounts of energy are produced in the cores of stars. Enormous temperatures and pressures cause nuclear fusion – atoms of hydrogen join together to form helium
Solar system	A star and the objects that orbit it	Planets, moons and asteroids (smaller lumps of rock) orbit our Sun. Comets also orbit the Sun, but usually have a long orbit cycle
Sun	The star at the centre of our solar system	It is estimated that the Sun is approximately 4·6 billion years old and that it is halfway through its life
Galaxy	A collection of stars	Astronomers have identified three main types of galaxy: spiral, elliptical and irregular
Universe	All the energy and matter (galaxies, stars, planets, interstellar gas and space) in existence	The current theory for the formation of the universe is the Big Bang model. The Big Bang model estimates the age of the universe to be approximately 13·7 billion years.
Dark matter	Scientists believe, from theoretical calculations, that there is a huge amount of matter in the universe that cannot be seen	Dark matter has so far not been detected by scientific instruments
Dark energy	Scientists believe that there is a source of energy that keeps the universe expanding, but they have not been able to identify it	Scientific instruments have so far been unable to detect dark energy

The light year

The **light year** is a unit of distance used in astronomy. A light year is defined as the distance that light travels in one year.

$d = vt$ 1 light year = $3 \times 10^8 \times 365{\cdot}25 \times 24 \times 60 \times 60 = 9{\cdot}47 \times 10^{15}$ m

Note that, for this calculation, 365·25 days, the average number of days in a year, is used.

When the huge distances connected with stars and galaxies are calculated in metres, it is difficult to make comparisons because all the results expressed in metres are so large. Light years are easier to interpret because they are a very large unit of distance.

To reach the Earth, light takes approximately:
- eight minutes from the Sun
- 4·3 years from the next nearest star
- 100 000 years from the opposite edge of our galaxy.

The Andromeda galaxy, our nearest galaxy, is approximately 2·5 million light years from Earth.

The distances between stars are huge.

Scale of the solar system and universe measured in light years

The diameter of our solar system measured across the outermost planetary diameter is approximately 0·0013 light years or $1·23 \times 10^{13}$ m.

The Earth is approximately 46·5 billion light years from the most distant edge of the observable universe.

The Andromeda galaxy.

Exoplanets

Interest in exoplanets has increased since the first discovery of an exoplanet orbiting a distant star.

Scientists think that some exoplanets may have suitable conditions that might be able to support life in some form.

Scientists refer to the **habitable zone** around a star. This is the zone in which a planet would have the appropriate conditions (such as surface pressure and temperature) to allow liquid water to exist. Water is essential to all life on Earth. The Earth is used here as a model to describe the requirements for life to exist on other planets.

The solar system.

The table gives the basic requirements for a planet to be in a potentially habitable zone around a star.

Requirement	Additional information
Liquid water	Wide expanses of water in liquid form should be available
Suitable temperature	The planet would need a surface temperature within the limits required for life to exist (i.e. no huge extremes in surface temperatures)
Oxygen	This is required by humans and most other species on Earth
Food	The planet would need conditions that allowed food to be produced

The enormous distances to other stars that support exoplanets means that travel to an exoplanet is not currently possible. The fastest speed possible in the universe is the speed of light in a vacuum (3×10^8 m s^{-1}). Even if a spacecraft could travel close to this speed, it would take years to reach the planet. Current space travel technology cannot meet these requirements.

Distant galaxies.

THINGS TO DO AND THINK ABOUT

1. Research the solar system and construct a table for the Sun and planets that lists:
   - the distance of each planet from the Sun;
   - the time taken for the planet to go round the Sun;
   - the time taken for each planet to complete one spin on its axis spin (in Earth-days or Earth-hours);
   - the gravitational field strength of each planet (N kg^{-1}).

2. Carry out research into the standard Big Bang model. Debate with others whether the huge cost of researching the Big Bang is worthwhile. For example, you could debate whether this knowledge has a real impact on our society or simply adds complicated information that is not useful.

DON'T FORGET

A light year is a measure of distance and is the distance travelled by light in one year

DON'T FORGET

Be able to describe the different types of objects in the universe.

EXTENDED QUESTIONS 1

SPEED AND ACCELERATION

1. Explain what is meant by 'average speed'.

2. A student travelled 750 m on a bicycle in 150 s. Calculate the student's average speed.

3. Copy and complete the following table.

	Average speed (m s⁻¹)	Distance (m)	Time (s)
(a)		420	35
(b)		86·4	5·52
(c)	14·8		8·6
(d)	3×10^8		2×10^{-3}
(e)	7·2	5·4	
(f)	126·4	19·2	

4. Between the start and finish of a race, an athlete ran at an average speed of 6·4 m s⁻¹. The athlete took 781·25 s to complete the race. Calculate the distance run by the athlete.

5. A bus travelled a distance of 24 km in 12 minutes. Calculate the average speed of the bus in m s⁻¹.

6. While cycling in the country, a cyclist reached the top of a hill and started to 'freewheel' downhill. The cyclist crouched down on the bike.

   (a) State the effect that this had on the frictional force acting on the cyclist.
   (b) Eventually, the downhill and frictional forces acting on the cyclist were balanced. State the effect this had on the speed of the cyclist.

7. At a school sports track, a student used a stopwatch to measure the time taken for a runner to complete a race. State what other measurement should be have been made to determine the average speed of the runner.
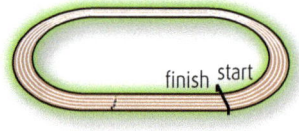

8. A footballer scored a penalty kick at a football match.

   The football travels a distance of 12 m from the penalty spot to the goal nets at an average speed of 25 m s⁻¹.

   (a) Calculate the time taken for the ball to reach the nets.
   (b) A speed–time graph showing how the speed

of the ball changes while in contact with the footballer's boot is shown. Calculate the acceleration of the ball while it is being kicked.
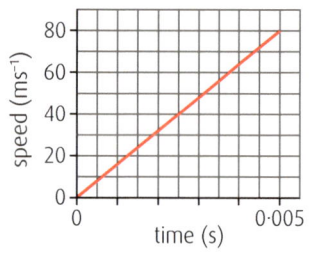

9. The distances and times of a train journey from Dundee to Stonehaven are shown in the table. Calculate the average speed of the train in m s⁻¹ for the whole journey.

Dundee → Arbroath	Arbroath	Arbroath → Montrose	Montrose	Montrose → Stonehaven
21·1 km	Waiting	19·3 km	Waiting	32 km
16 minutes	2 minutes	17 minutes	3 minutes	21 minutes

10. Satellite tracking is used to monitor the movements of whales in oceans as they migrate. One whale travelled 4 924 800 m in 15 days. Calculate the average speed the whale during its journey:

    (a) in km h⁻¹
    (b) in m s⁻¹

11. Explain what is meant by 'instantaneous speed'.

12. A trolley is released on a slope and travels through a light gate on the slope in a time of 0·16 s. The width of the card is 0·06 m.
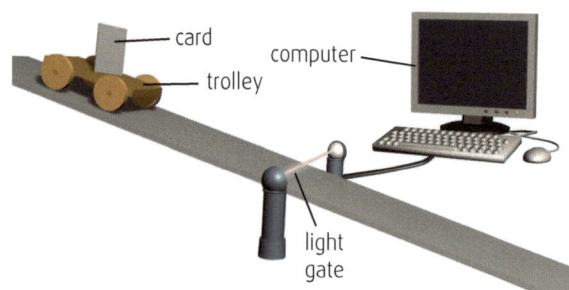

Calculate the instantaneous speed of the trolley as it passes through the light gate.

13. A bottle on a conveyer belt passes through a light gate in 0·15 s. The width of the bottle is 9 cm. Calculate the instantaneous speed of the bottle as it passes through.

14. Which of the following speed–time graphs represents the motion of a vehicle:

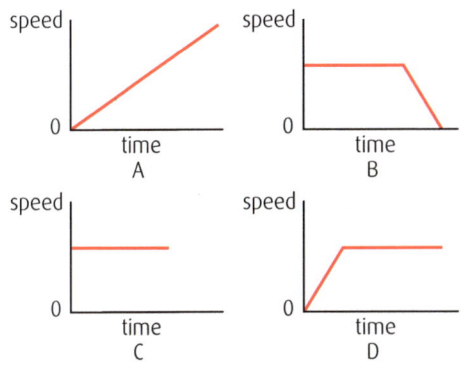

(a) moving at constant speed
(b) accelerating from rest
(c) moving at constant speed then decelerating to rest?

15. The speed–time graph of a car for the first 10 s after it leaves traffic lights is shown. Calculate the distance travelled by the car during this time.

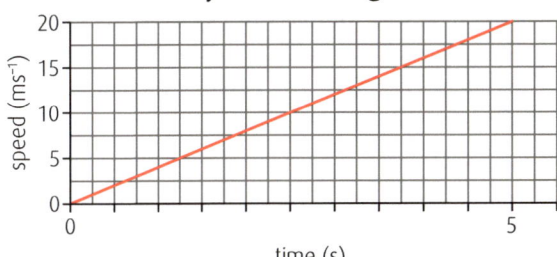

16. The speed–time graph of part of the journey of an aeroplane is shown. Calculate the distance travelled by the aeroplane during this time.

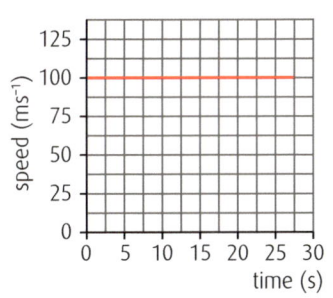

17. The speed–time graph of a tram moving from one stop to the next is shown. Calculate the distance travelled by the tram between the stops.

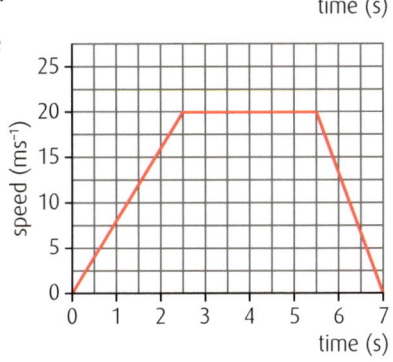

18. A bus changes speed by 8·5 m s⁻¹ in 2·6 s. Calculate the acceleration of the bus.

19. A fighter jet accelerates at 23·4 m s⁻². The jet's speed changes by 128·7 m s⁻¹. Calculate the time taken for this acceleration.

20. A car accelerates at 6·2 m s⁻² for 2·5 s. Calculate the change in speed of the car.

21. A jumping insect had a change of speed of 1·4 m s⁻¹ in a time of 0·02 s. Calculate the insect's acceleration.

22. Copy and complete the table.

	Acceleration (m s⁻²)	Change in speed (m s⁻¹)	Time (s)
(a)		35	5
(b)		50·6	5·5
(c)	12·6		14·5
(d)	3 × 10⁻³		2 × 10⁻³
(e)	14·4	10·8	
(f)	60·2	451·5	

23. A trolley is released at the top of a ramp. A card attached to the trolley passes through two light gates. At each light gate, a computer calculates the trolley speed.

The computer then calculates the change in speed of the trolley between the light gates. A stopwatch is used to measure the time taken for the trolley to travel between the two gates.

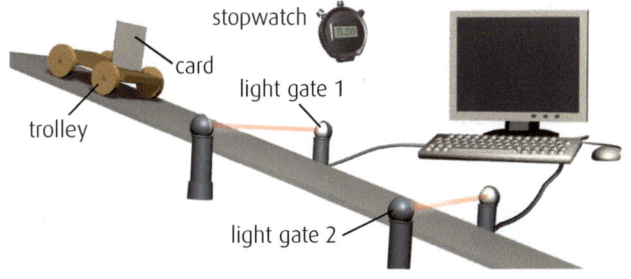

The experiment was repeated several times. Each time, the light gates were placed at different positions on the ramp.

Copy and complete the table of results.

Run	Change in speed (m s⁻¹)	Time (s)	Acceleration (m s⁻²)
1	0·21	0·25	
2	0·43	0·50	
3	0·63	0·75	

EXTENDED QUESTIONS 2

RELATIONSHIP BETWEEN FORCES, MOTION AND ENERGY

1. State three effects that a force can have on an object.

2. A skier takes part in a downhill racing contest. Suggest two ways that the skier could reduce friction in order to go faster.

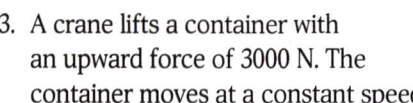

3. A crane lifts a container with an upward force of 3000 N. The container moves at a constant speed.
   (a) State the weight of the container.
   (b) Explain your answer.

4. A rowing team exerts a total forward force of 750 N on their boat.

   The frictional force acting on the boat is 750 N in the opposite direction. Describe the movement of the boat.

5. Which of the objects below have an unbalanced force acting on them?

   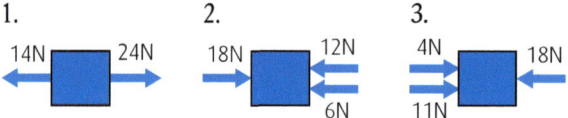

6. Calculate the unbalanced force required to accelerate a 750 kg car at 2·5 m s^{-2}.

7. Copy and complete the table.

	Unbalanced force (N)	Mass (kg)	Acceleration (m s^{-2})
(a)		18	4
(b)		36·6	2·2
(c)	24·6		1·23
(d)	2 × 10^3		4 × 10^{-2}
(e)	16·2	4·05	
(f)	8800	4000	

8. Calculate the weight of an elephant of mass 5350 kg.

9. At an airport, a packed suitcase has a weight of 240·1 N. Determine whether or not the suitcase is below the airline's 23 kg baggage limit.

10. Copy and complete the table. Note you will need to look back and find the gravitational acceleration for each planet.

Planet	Mass of object (kg)	Weight of object (N)
Earth	5·6	
Venus		22·25
Mars	8·2	
Saturn		1·8 × 10^3
Mercury	16·5	
Neptune		4400

11. An astronaut's mass on Earth is 65 kg.
    (a) What is the weight of the astronaut: (i) on Earth and (ii) on the Moon?
    (b) What is the mass of the astronaut on the Moon?

12. In 2014, a spacecraft called the Rosetta successfully landed a probe of mass 100 kg on Comet 67P. The gravitational field strength of the comet was estimated at 5·24 × 10^{-5} N kg^{-1}. Calculate the weight of the probe on the surface of the comet.

SATELLITES

1. Satellites can be used to house instruments that detect and gather data from space. Explain why it is necessary to have these instruments on the satellite and not based on the surface of the Earth.

2. There are many satellites in orbit above the Earth with a variety of uses. Make a list of at least three different applications of orbiting satellites.

3. Explain what is meant by the term 'geostationary satellite'.

4. The period of a satellite is the time taken for it to complete a rotation around the Earth. Which of the following statements about the satellite's orbit is correct?

   A The closer a satellite is to the Earth, the longer the period.

   B The period of a satellite does not depend on its distance from Earth.

   C The period of a satellite depends on the mass of the satellite.

   D The further away a satellite is from Earth, the longer its period.

5. The International Space Station has a period of 93 minutes and an orbital height of 400 km. The geostationary weather satellite GOES-15 has

a period of 1436 minutes and an orbital height of 35 780 km.

Which of the following is the period of a satellite that has an orbital height of 20 000 km?

A 82 minutes
B 93 minutes
C 720 minutes
D 1436 minutes
E 1850 minutes

6. A curved reflector is used to receive radio signals from satellites.

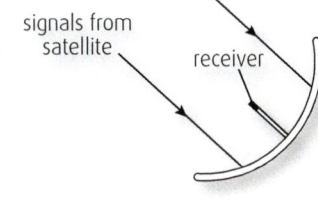

signals from satellite

receiver

(a) Copy and complete the diagram to show how the signals reach the receiver of the dish aerial.

(b) What effect would a larger diameter dish have on the signal detected at the receiver?

7. A communications satellite is in orbit 36 000 km above the Earth.

(a) Calculate the time taken for a signal sent from Earth to reach the satellite.

(b) Explain why the time taken causes a delay when two people have a phone call using a communication satellite.

8. A signal sent from a geostationary satellite takes 0·13 s to reach a receiver on Earth. Calculate the distance of the satellite from the receiver.

9. In 2014 the Rosetta spacecraft reached comet CP67, a distance of $4·05 \times 10^{11}$ m from Earth. Calculate the time taken for a signal sent from Earth to reach the spacecraft when it was next to the comet.

COSMOLOGY

1. List the following in order of increasing size: solar system – galaxy – universe – planet.

2. What is the definition of a light year?

3. Convert one light year into metres.

4. It takes light from the Sun eight minutes to reach Earth. Calculate the distance from the Sun to the Earth in metres.

5. One light year is equivalent to $9·47 \times 10^{15}$ m. The star Polaris is estimated to be 323 light years from Earth. Calculate the distance of Polaris from the Earth in metres.

6. According to the Big Bang model theory, what is the current estimated age of the universe (in billions of years)?

7. What is the definition of a planet?

8. What is the definition of a moon?

9. What is the definition of a galaxy?

10. The table gives some information about planets that orbit our Sun.

Planet	Estimated distance from the Sun ($\times 10^9$ m)	Period of orbit (Earth-days)	Mass (Earth-masses)
Earth	150	365	1
Jupiter	780	4332	318
Mars	228	687	0·11
Mercury	57	88	0·06
Saturn	1430	10 760	95
Venus	108		0·82

(a) Suggest an approximate value, in Earth-days, for the period of Venus.

(b) Calculate the time taken for light from the Sun to reach Mars (speed of light in vacuum = 3×10^8 m s^{-1}).

11. The table gives information about some of the planets in our solar system.

(i) Which two planets have the same length of day?

(ii) On which planets will a 4 kg mass have the same weight?

(iii) Which planet has the longest solar orbit time?

Planet	Mercury	Venus	Earth	Mars
Distance from the Sun (million km)	58	110	150	228
Time to orbit the Sun once (Earth-years)	0·24	0·6	1	1·9
Time for one complete spin (in Earth-days or Earth-hours)	59 days	243 days	24 hours	25 hours
Gravitational field strength (N kg^{-1})	3·7	8·9	9·8	3·7

Planet	Jupiter	Saturn	Neptune
Distance from the Sun (million km)	780	1430	4500
Time to orbit the Sun once (Earth-years)	12	29	165
Time for one complete spin (in Earth-days or Earth-hours)	10 hours	10 hours	16 hours
Gravitational field strength (N kg^{-1})	23	9·0	11

12. More than 1800 exoplanets have now been discovered.

(a) Explain what is meant by an exoplanet.

(b) State two basic requirements for a planet to be habitable.

COURSE ASSESSMENT

OVERVIEW OF THE ASSESSMENT

You are not required to sit a final exam at the end of your National 4 Physics studies to achieve a course award. Instead of a final exam, you are required to pass four assessment tasks. These four tasks will be taken in your school or college and they will be marked by your teacher or lecturer on a pass or fail basis.

WHAT ARE THE ASSESSMENT TASKS?

- **Assessment task 1.** Preparation of a scientific report on a physics experiment or practical investigation.
- **Assessment task 2.** A short scientific report based on your research of a physics topic.
- **Assessment task 3.** A set of questions on the physics you have covered during the course. You will complete this assessment task for each unit. In this book, the **Quick questions** at the end of each page and the **Extended questions** at the end of each unit provide an opportunity to practise similar questions to those in the assessment.
- **Assessment task 4.** This task is called the **Added Value Unit** or **Physics Assignment**. Detailed information on this can be found on page 92 of this Study Guide.

ASSESSMENT TASK ONE

To be successful in this assessment task you must plan and carry out an experiment or practical investigation and then prepare a scientific report about the work undertaken. Your report must cover the following areas.

Planning

This section of the report must include:

(i) The **aim** of the experiment: what you are trying to find out. Some examples of suitable aims are:

- To find out how the sound from a loudspeaker detected by a sound level meter depends on the distance of the meter from the loudspeaker.
- To find out how wind speed affects the output power of a wind turbine.
- To find out how the speed of a trolley when it reaches the bottom of a slope is affected by the angle of the slope.

(ii) **Variables:** the variable that you change during the experiment (for example, the temperature of water in a beaker or the voltage across a lamp) is known as the **independent variable**.

The variable that is affected by the changes you made to the independent variable is known as the **dependent variable**.

For example, in an experiment to investigate a thermistor, a student changes the temperature (independent variable) of the thermistor. The resistance (dependent variable) of the thermistor will then change.

You also need to state the variable(s) that needs to be kept constant in order to keep the experiment fair.

(iii) The **measurements** or **observations** to be made.

(iv) The resources (**apparatus**) to be used. A labelled diagram often helps to make the experiment clear.

DON'T FORGET

A variable in an experiment is something that can be altered – for example, the distance from a meter, the voltage across a light bulb, the size of the force pulling a trolley and the mass of the trolley are all variables. For an experiment to be valid, only one variable should be changed by the student carrying out the experiment. The other variables must remain unchanged.

(v) How the experiment will be carried out – the **procedure**. This must be written in such a way that another National 4 pupil could use the information to repeat the experiment.

(vi) Any **safety** precautions needed. This could include the use of safety glasses, not allowing trolleys to fall to the floor when moving down a ramp or not exceeding the recommended voltage for the lamps used.

Following procedures safely

Your teacher will check that you carry out the experiment in a safe manner.

Making and recording observations/measurements accurately

Always record the results from an experiment in a table.

Presenting results in an appropriate manner

Some things you might include are:

- tables with appropriate headings and units;
- calculations of average results from repeated experiments;
- bar charts or line graphs – charts or graphs should have appropriate scales, labels and units, and the bars or points should be plotted correctly.

Drawing valid conclusions

Your conclusion must relate to your aim.

The valid conclusions for the aims set out could be written as:

- the power output of the turbine increased as the wind speed increased;
- the resistance of the thermistor decreased when its temperature was increased;
- the trolley speed at the bottom of the slope increased when the angle of the slope was increased.

Evaluating the procedure

In this final section of the report, you must suggest something that would improve the experiment.

ASSESSMENT TASK TWO

To be successful in this assessment, you must select and research an application of physics from one of the Key Areas you have studied. The report should show the impact that the application has had on the environment or society.

When you have completed your report, you should read over it and make sure that you have:

- stated the application of physics involved;
- used appropriate physics knowledge to describe the application – for example, if your application was the use of ultraviolet radiation, you could state that ultraviolet radiation can be used by dentists to harden fillings.
- stated how the environment and/or society has been affected by the application;
- used appropriate physics knowledge to describe its effect – for example, if your application was the use of seatbelts in cars, you could state that wearing seatbelts has a positive impact on society because this reduces the number of serious injuries in car crashes.

DON'T FORGET

Remember to ask your teacher for the candidate guides for the assessment tasks.

ASSESSMENT TASK THREE

This consists of sets of questions on the physics you have covered in the N4 Physics course. You will complete this assessment task for each unit. In this book, the **Quick questions** at the end of each page and the **Extended questions** at the end of each unit provide an opportunity to practise similar questions to those in the assessment.

ASSESSMENT TASK FOUR

The **Added Value Unit (Physics Assignment)** at N4 takes the place of a final examination and is marked as a pass or fail. It covers the following assessment standards:

1.1. Choosing, with justification, a relevant issue in physics

1.2. Researching the issue

1.3 Presenting appropriate information/data

1.4 Explaining the impact, in terms of the physics involved

1.5. Communicating the findings of the investigation.

If you fail to pass one or more of these assessment standards, **don't panic**, because you are given **one** opportunity to resit the assessment.

You will apply your skills, subject knowledge and understanding to investigate a topical issue in physics and its impact on the environment and/or society.

There are two stages to the Added Value Unit process.

STAGE ONE: THE RESEARCH STAGE

First, you must select a topical issue on which to base your Added Value Unit.

Next, you will gather information from a variety of sources (such as books, the internet, class notes or scientific journals). This will allow you to create a candidate's log or journal.

Your teacher will help you with this stage, although the research must be your own work and you should record the sources for any information that you gather.

If you carry out an experiment as part of the research stage, you may wish to include this work in your assignment.

STAGE TWO: THE COMMUNICATION STAGE

In this stage, you will produce your final assignment.

During the second stage of the assignment, you will select, use and record, as a minimum, at least two appropriate sources from your research that will be included in your final communication.

You must select, process and present the information and/or data relating to your chosen topical issue that you gathered in stage one.

You can choose how you communicate your findings and this may include one or more of the following:

- a written or word-processed report (200–400 words is the suggested length);
- a presentation, oral or digital, accompanied by supplementary material such as presentation notes or presentation slides with notes;
- an information booklet/leaflet;
- an information poster (this poster must include annotated notes).

THINGS TO DO AND THINK ABOUT

Be sure to make good use of this book and all of your course materials as you prepare for your National 4 Physics assessments – you have the skills and the tools necessary to succeed and, as long as you do your best, you will achieve a qualification of which you can be proud.

DON'T FORGET

Your teacher will support you throughout your Added Value Unit.

JUST A WEE NOTE

The impact of the topical issue on the environment and/or society can be positive and/or negative.

DON'T FORGET

Group work is allowed for this stage, but you will need to show you have actively participated in any group work.

DON'T FORGET

Remember that Wikipedia may not be a reliable source.

DON'T FORGET

Remember to include the title and aim of any experiments.

JUST A WEE NOTE

This will take place under controlled conditions – this means your teacher or lecturer will be present.

ANSWERS

ANSWERS TO **THINGS TO DO AND THINK ABOUT** AND **QUICK QUESTIONS**

ELECTRICITY AND ENERGY

Generation of electricity: where our electricity comes from 1

Things to do and think about
1. (a) (i) Chemical energy → heat energy;
   (ii) kinetic energy → electrical energy.
   (b) Carbon dioxide.
   (c) Carbon dioxide emissions are thought to contribute to global warming.
2. (a) (i) Gravitational potential energy → kinetic energy;
   (ii) kinetic energy → electrical energy.
   (b) Hydroelectric power stations use renewable sources of input energy.
   (c) Hydroelectric power stations are found in mountainous areas.
   (d) Pumped storage reservoirs can be refilled for use at peak times of energy demand.

Generation of electricity: where our electricity comes from 2

Things to do and think about
1. (a) Generator; (b) reactor.
2. Uranium nuclei are split into smaller nuclei with a release of energy, which results in the production of heat.
3. Far less pollution from atmospheric emissions of greenhouse gases.
4. Nuclear waste has to be carefully stored for a long time before it is safe; costly to dismantle power stations at the end of their life due to contamination with radioactive substances.

Generation of electricity: where our electricity comes from 3

Quick questions
1. A voltage is produced when a magnet is moved inside a coil of wire or a coil of wire is moved around a magnet.
2. The voltage produced can be increased by: increasing the strength of the magnetic field; increasing the speed of movement of the magnet or coil; or by increasing the number of turns of wire on the coil.

Generation of electricity: where our electricity comes from 4

Quick questions
1. (a) Renewable energy source, non-polluting.
   (b) Only available in the daytime, not efficient in more northerly countries.

Distribution of electricity

Quick questions
1. Transmission lines (overhead cables) are used to transport electrical energy.
2. Transformers are used to increase and decrease the voltage for transmission, which reduces energy losses.
3. 30 V

Electrical power 1

Quick questions
1. The unit of electrical power is the watt (W).
2. Domestic appliances used for heating usually have the largest power ratings.

Things to do and think about
1. $E = Pt = 3000 \times 2 \times 60 = 360\ 000\ \text{J}$
2. $P = \dfrac{E}{t} = \dfrac{1\ 008\ 000}{2 \times 60 \times 60} = 140\ \text{W}$

Electrical power 2

Quick questions
1. 39%
2. 27%

Electromagnetism

Quick questions 1
1. A magnetic field pattern can be represented by drawing magnetic field lines.
2. The direction of magnetic field lines is north to south.
3. The magnetic field is strongest where the magnetic field lines are closest together.

Quick questions 2
1. A magnetic field is produced.
2. A magnetic field is produced.
3. An electromagnet consists of a coil of wire wound on an iron core.
4. Increasing the number of turns in the coil will increase the magnetic field.

Things to do and think about
2. (i)   (ii)   (iii)

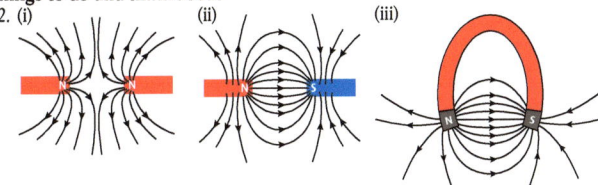

Practical electrical and electronic circuits 1

Quick questions 1
1. The movement of electrons can cause a current to flow in a wire.
2. An ammeter is placed in series with the lamp to measure the current through it.
3. Voltage is a measure of the energy required to move a charge through a component.
4. A voltmeter is placed in parallel with the lamp to measure the voltage across it.

Quick questions 2
1.

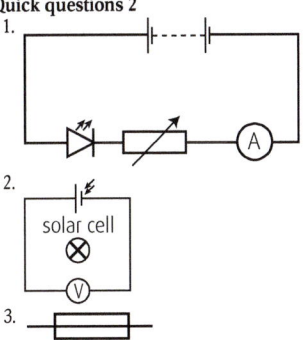

2.

3.

4. A transistor is a semiconductor device that acts as an electronic switch or amplifier.

Practical electrical and electronic circuits 2

Quick questions 1
1. Thermistor.
2. Light-dependent resistor.
3. Capacitor.

Quick questions 2
1. Bedside lamp.
2. Stair lights.
3. Less wiring is required and it is cheaper.

Practical electrical and electronic circuits 3

Quick questions
1. An OR logic gate.
2. An AND logic gate.

Things to do and think about
Example: microwave oven.
Order of processes in microwave oven:
  • has start button been pressed?
  • has time of cooking been selected?
  • is machine door closed?
  • if so, start machine (if door is opened when cooking is taking place, then stop heating);
  • when cooking time is complete, stop microwave process;
  • sound warning signal to indicate completion.

Practical electrical and electronic circuits 4

Things to do and think about
1. 2 A
2. 10 V

Resistance and Ohm's law

Quick questions
1. The current decreases.
2. The resistance of a conductor depends on its length, cross-sectional area and type of material.
3. The resistance of a wire increases when its length is increased.
4. The resistance of a wire decreases when its cross-sectional area is increased.
5. It may overheat and melt.
6. Copper is both cheap and a good conductor.

Things to do and think about
1. $V = IR$
2. 2·9 Ω
3. 1·3 A
4. 4·5 V

Gas laws and the kinetic model

Things to do and think about
1. The person's weight is spread over a large area and so the pressure on the snow is reduced.
2. Pressure exerted by the elephant = 70 000 Pa

ANSWERS

ANSWERS TO **THINGS TO DO AND THINK ABOUT** AND **QUICK QUESTIONS**

Gas pressure
Things to do and think about
1. The gas molecules increase their speed.
2. The gas molecules exert a force on the walls of the container.
3. The volume will decrease.
4. The pressure will decrease.
5. The volume will increase.

WAVES AND RADIATION
Wave characteristics 1
Things to do and think about
1. Energy.
2. Longitudinal and transverse.
3. At right angles to the direction of travel.
4. In the same direction as the direction of travel.
5. Transverse waves – light; longitudinal waves – sound waves.

Wave characteristics 2
Quick questions
1. $\lambda = 8.0$ cm, $a = 2$ cm
2. 0·3 Hz
3. (a) 5 Hz; (b) 5 V

Things to do and think about
1. 0·3 m s⁻¹
2. 600 Hz
3. 6·5 m
4. 2·5 m s⁻¹
5. 100 s
6. 2720 m

Sound 1
Things to do and think about
(a) C
(b) A

Sound 2
Things to do and think about
1. 320 m s⁻¹
2. 333 m s⁻¹
3. 816 m

Sound 3
Things to do and think about
1. Decibels (dB)
2. 90 dB
3. Pneumatic drill.
4. Sound insulation materials are used in construction.
5. 5933 s

Electromagnetic spectrum 1
Things to do and think about
1. 300 000 000 m s⁻¹
2. B = X-rays; D = visible light.
3. An aerial.
4. Gamma rays.
5. Radio waves.

Electromagnetic spectrum 2
Things to do and think about
1. (a) Concave. (b) Convex.
2. A Geiger–Müller tube can be used to detect gamma rays.
3. Overexposure to X-rays could cause cancer.
4. Gamma radiation is used to sterilise medical instruments.
5. A black bulb thermometer can detect infra-red radiation.
6. Gamma rays, X-rays and most ultraviolet radiation are filtered by the Earth's atmosphere.
7. Radioactive waste requires expensive, long-term storage until it becomes safe.
8. Ultraviolet radiation is used to harden dental fillings.

Nuclear radiation 1
Quick questions
1. Protons, neutrons.
2. Protons are positive; neutrons have no charge.
3. Electrons are negative.
4. Some rocks, e.g. granite.
5. Nuclear reactors.
6. Alpha (α), beta (β) and gamma (γ) radiation.
7. Alpha radiation.
8. Gamma radiation.

Nuclear radiation 2
Things to do and think about
1. Radiation is emitted when an atom decays.
2. The activity of a radioactive substance is the number of nuclei decaying every second.

3. The half-life of a radioactive element is the time required for the activity of the element to decrease to half of its original value.

Applications and uses of nuclear radiation
Quick questions
1. Radioactive tracers are used in medical diagnosis.
2. Radiation is used to treat cancer.
3. Radiation is used to check the thickness of newspaper.

DYNAMICS AND SPACE
Speed and acceleration 1
Things to do and think about
1. 6 m s⁻¹
2. 2520 s, 0·7 hours
3. 964800 m = 964·8 km
4. 30 m s⁻¹
5. Average speed is measured over the total distance; instantaneous speed is measured over a short period of time.

Speed and acceleration 2
Things to do and think about
1. (a) E; (b) D; (c) B; (d) A; (e) C
2. (a) 9 s; (b) 14 s; (c) 5 s

Speed and acceleration 3
Quick question
1. 2 m

Things to do and think about

1.

	Δv (m s⁻¹)	t (s)	a (m s⁻²)
(a)	14	0·7	20
(b)	0·05	2·5	0·02
(c)	378	1·4	270

Relationship between forces, motion and energy 1
Things to do and think about
1. The force can change the object's speed, shape and direction.
2. The forces are balanced.
3. It remains stationary.
4. It moves at a constant speed.

Relationship between forces, motion and energy 2
Things to do and think about
1. The object accelerates.
2. Its mass and the unbalanced force acting on it.
3. $F = ma$
4. 2025 N
5. 4·8 m s⁻²
6. 8500 kg

Relationship between forces, motion and energy 3
Things to do and think about
The Earth's gravitational field strength reduces as the distance above the Earth's surface increases. Objects at higher altitudes have a slightly smaller weight.

Cosmology
Things to do and think about

1.

Planet	Distance from the Sun (million km)	Time to taken to orbit the Sun (Earth-years)	Time taken for one rotation (Earth-hours or Earth-days)	Gravitational field strength (N kg⁻¹)
Mercury	58	0·24	59 days	3·7
Venus	110	0·6	243 days	8·9
Earth	150	1	1 day	9·8
Mars	228	1·9	25 hours	3·7
Jupiter	780	12	10 hours	23
Saturn	1430	30	10 hours	9·0
Uranus	2877	84	18 hours	8·7
Neptune	4500	165	19 hours	11

ANSWERS TO **EXTENDED QUESTIONS**

ELECTRICITY AND ENERGY
Generation of electricity

1. (a) (i) Chemical energy → heat energy;
   (ii) kinetic energy → electrical energy.
   (b) High pressure steam is required to turn the turbines attached to the generator.
2. (a) Nuclear waste; (b) The waste materials are radioactive and very dangerous; they require long-term, safe storage.
3. (a) (i) Gravitational potential energy → kinetic energy;
   (ii) kinetic energy → electrical energy.
4. Number of wind farms required = $\dfrac{\text{power output of Longannet}}{\text{power output of one windfarm}}$
   $= \dfrac{2\cdot3}{0\cdot124} = 18\cdot5$ so **19** windfarms would be required.
5. Unlike an ordinary hydroelectric power station, a pumped storage power station can pump water from ground level to quickly refill a reservoir to allow fast access to electrical energy when required.
6. In nuclear reactors, a **uranium nucleus** is bombarded by a **neutron** and splits into two smaller nuclei and more neutrons. This splitting is called **fission**. There is more kinetic energy after the fission than before and this is how the **heat energy** is produced. The extra neutrons released go on to produce more fission and more heat. This is called a **chain reaction**.
7. Number of kilograms of coal required = $\dfrac{\text{energy from 1 kg uranium}}{\text{energy from 1 kg coal}}$
   $= \dfrac{4\cdot5 \times 10^{12}}{3\cdot0 \times 10^{7}}$
   $= 1\cdot5 \times 10^{5}$ kg of coal required.

8.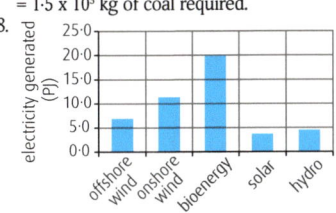

9.

Source of energy	Advantages	Disadvantages
Wind	Will not run out, non-polluting during use, free at source	Wind supply is variable
Solar	Will not run out, non-polluting during use, free at source	Poor output in cooler countries, requires large area of solar cells to produce significant output
Coal	Still has a long-term supply in some countries	Causes atmospheric pollution, including greenhouse gases, dirty and can cause damage to surroundings
Nuclear	Only small quantities are required, long lifetime of supply	Waste material remains hazardous for long periods, accidents can cause severe pollution
Hydro	Will not run out, non-polluting during use, free at source	Requires mountainous terrain

10. (a) It is possible to create a **voltage** and current in a **coil of wire** by the **movement** of a **magnet** towards (or away from) the coil. An **iron core** inside the coil increases the size of the voltage induced.
    (b) More coils or moving the magnet more quickly will increase the induced voltage.
11. Solar cells can be placed on the roofs of individual houses to reduce the electrical energy required from the National Grid.
12. (a) The current in the wires causes heat energy to be produced – this is the main cause of energy loss.
    (b) One way of reducing this loss of energy is to use transformers to transfer energy from the power station across the country at very high voltages.

13.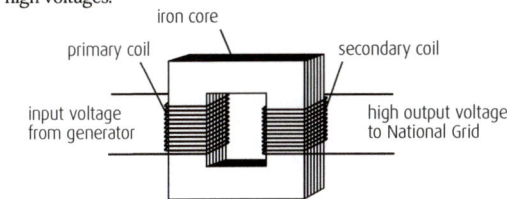
    iron core
    primary coil   secondary coil
    input voltage   high output voltage
    from generator   to National Grid

14. $\dfrac{V_s}{V_p} = \dfrac{n_s}{n_p}$
    $\dfrac{750}{11\,000} = \dfrac{n_s}{88\,000}$
    $n_s = 6000$ turns

15. $\dfrac{V_s}{V_p} = \dfrac{n_s}{n_p}$
    $\dfrac{V_s}{230} = \dfrac{504}{6440}$
    $V_s = 18$ V

16. $\dfrac{V_s}{V_p} = \dfrac{n_s}{n_p}$
    $\dfrac{V_s}{20} = \dfrac{500}{200}$
    $V_s = 50$ V

17. (a) Transformers are used to step-up the voltage from power stations to allow electrical energy to be transmitted across the country at very high voltages to reduce energy loss. Step-down transformers reduce this voltage to a safer level for use in towns.
    (b) (i) 132 000 V
        (ii) 11 000 V
        (iii) $\dfrac{V_s}{V_p} = \dfrac{n_s}{n_p}$
            $\dfrac{11\,0000}{132\,000} = \dfrac{n_s}{9000}$
            $n_s = 750$ turns

Electrical power

1. The power rating of an appliance is the number of joules of energy used per second.
2. (a) 1600 J; (b) heating only the required amount of water will use less energy.
3. The kettle would transfer the most power.
4. $P = \dfrac{E}{t}$
   $E = 12 \times 60 = 720$ J
5. $P = \dfrac{E}{t}$
   $E = 900 \times 180 = 162\,000$ J
6. In power stations, energy is lost during the conversion process of one form of energy to another until electrical energy is produced.
7. % Efficiency $= \dfrac{\text{Useful } E_O}{E_I} \times 100$
   % Efficiency $= \dfrac{570\cdot4 \times 4 \times 10^{6}}{620 \times 10^{6}} \times 100$
   Efficiency $= 92\%$
8. % Efficiency $= \dfrac{\text{Useful } E_O}{E_I} \times 100$
   % Efficiency $= \dfrac{49\cdot98 \times 10^{6}}{58\cdot8 \times 10^{6}} \times 100$
   Efficiency $= 85\%$
9. % Efficiency $= \dfrac{\text{Useful } E_O}{E_I} \times 100$
   % Efficiency $= \dfrac{44\,955}{60\,750} \times 100$
   Efficiency $= 74\%$

Electromagnetism

1.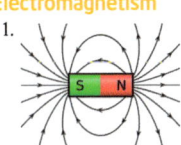

2. (a) Winding more turns of wire on the nail would increase the strength of the electromagnet.
   (b) The pins would fall because the magnetic field had been removed.
3. Loudspeakers, simple d.c. motors.

Practical electrical and electronic circuits

1. Electrons.
2.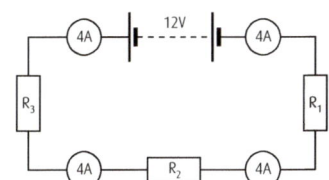

3.

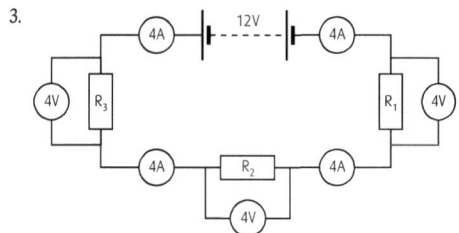

4.

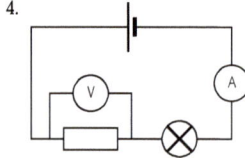

5. (i) ——☐—— (ii) ——☐—— (iii) ——☐——

6. Any value between 125 and 200 Ω.

7. (a) Total energy in one year = $18 \times 5.2 \times 10^8 = 9.36 \times 10^9$ J

(b) Total cost of energy in one year = cost per unit × $\dfrac{\text{total energy in year}}{\text{energy per unit}}$

Total cost of energy in one year = $19.4 \times \dfrac{9.36 \times 10^9}{3.6 \times 10^6} = £504.40$

(c) The amount of sunlight varies from year to year, so the electrical energy generated the next year may be different.

8. $V = 0.03 \times 550 = 16.5$ V

9. $I =$ $v = 0.125$ A

10. $R = \dfrac{12}{2} = 6$ Ω

11. (a) When $V = 12$ V, $I = 5.0$ A
$V = IR$
$R = \dfrac{12}{5} = 2.4$ Ω

(b) $I = 11$ A, $R = 2.4$ Ω
$V = IR = 11 \times 2.4 = 26.4$ V

12. System 1 (OR gate).

13. Input devices: solar cell, light-dependent resistor.

14. Total resistance of 30 km cable = $0.3 \times 30 = 9$ Ω
$V = 18 \times 9 = 162$ V

15. Ammeter 2.

Gas laws and the kinetic model

1. (a) The gas pressure increases when the bottle is heated; (b) gas particles gain kinetic energy and move faster, hitting the container walls more often and with greater force, so the gas pressure increases.

2. When colder, the gas particles lose kinetic energy and move more slowly – they hit the tyre walls less often and with less force, so the pressure decreases.

3. (a) The balloon volume increases; (b) When warmer, the gas particles gain kinetic energy and move more quickly. They hit the container walls more often with greater force causing increased pressure, which causes the balloon volume to increase.

WAVES AND RADIATION

Wave characteristics

1. (i) Y = transverse; (ii) X = longitudinal.

2. $a = \dfrac{16}{2} = 8$ cm

3. $\lambda = \dfrac{30}{3} = 10$ cm

4. $\lambda = \dfrac{28}{4} = 7$ cm; $a = \dfrac{6}{2} = 3$ cm

5. $f = \dfrac{N}{t}$
$N = 2 \times 50 = 100$ Hz

6. $f = \dfrac{N}{t} = \dfrac{12}{36} = 0.3$ Hz

7. (a) $f = \dfrac{N}{t} = \dfrac{2}{0.8} = 2.5$ Hz; (b) $a = 10$ V

8. (i) Diagram 2 shows longitudinal waves (sound waves); (ii) diagram 1 shows transverse waves (water waves).

9. (a) P–S, Q–T, R–U.
(b) 5 Hz means that five complete waves are produced each second.

10. (a) $\lambda = \dfrac{1500}{80} = 18.75$ m

(b) $f = \dfrac{340}{1.7} = 200$ Hz

(c) $v = 950 \times 3.2 = 3040$ m s^{-1}

11. $f = \dfrac{340}{0.4} = 850$ Hz

12. $v = 12 \times 0.3 = 3.6$ m s^{-1}

13. $\lambda = \dfrac{5200}{650} = 8$ m

14. $v = \dfrac{24}{16} = 1.5$ m s^{-1}

15. $d = 4.2 \times 14 = 58.8$ m

16. $t = \dfrac{26.1}{5.8} = 4.5$ s

17. (a) $v = \dfrac{4.2}{16} = 0.26$ m s^{-1}

(b) $f = \dfrac{N}{t} = \dfrac{18}{24} = 0.75$ Hz

(c) $\lambda = \dfrac{0.26}{0.75} = 0.35$ m

18. (a) $v = \dfrac{d}{t} = \dfrac{1.4}{5.6} = 0.25$ m s^{-1}

(b) $f = \dfrac{N}{t} = \dfrac{28}{22.4} = 1.25$ Hz

(c) $\lambda = \dfrac{N}{f} = \dfrac{0.25}{1.5} = 0.2$ m s^{-1}

Sound

1. (a)

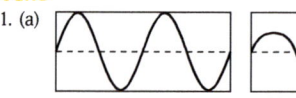

Figure 1   Figure 2

(b) The frequency control is used to adjust the number of waves produced each second.

2. (a) $v = \dfrac{d}{t} = \dfrac{1.16}{0.0008} = 1450$ m s^{-1}

(b) Place microphone 1 inside the water tank.

3. (a) 60 Hz.
(b) Elephant.
(c) Any frequency between 34 000 and 45 000 Hz.
(d) $\lambda = \dfrac{340}{50} = 6.8$ m

Electromagnetic spectrum

1. Overexposure to UV can cause skin cancer.

2. (a) Photographic film.
(b) Black bulb thermometer.

Nuclear radiation

1. Proton, positive charge; neutron, no charge; electron, negative charge.

2. A, electron; B, neutron; C, proton; and D, nucleus.

3. Artificial source, plutonium; natural source, radon gas.

4.

Alpha	α	Large positive nucleus
Beta	β	Fast-moving electron
Gamma	γ	Electromagnetic wave

5. Alpha radiation can only travel a few centimetres in air before being absorbed. Beta radiation can be absorbed by 3 or 4 cm of aluminium. Gamma rays require several centimetres of lead to absorb most of their energy.

6. Geiger–Müller tube and counter.

7. (a) Alpha radiation; (b) gamma radiation.

8. Atoms of radioactive elements emit radiation when they decay.

9. The activity of a radioactive substance is the number of atoms that decay each second. It is measured in becquerel (Bq).

10. The half-life of a radioactive element is the time taken for the **activity** of the element to decrease to one-half of its initial value.

11. (a) A: overexposure can cause cancer; (b) B: ultraviolet radiation; (c) C: treatment of muscle damage; (d) D: long distance radio communication.

12. (a) Arguments **for** using nuclear fuel as an energy source:
   - Nuclear power stations cause very little pollution of the atmosphere compared with power stations burning fossil fuels.
   - Many people believe that the carbon dioxide gas released by burning fossil fuels in power stations is a major cause of global warming. They say that global warming will cause irreversible damage to the Earth's